Y0-CCH-781

Seismic Applications of Homomorphic Signal Processing

PRENTICE-HALL SIGNAL PROCESSING SERIES

Alan V. Oppenheim, Editor

ANDREWS and HUNT *Digital Image Restoration*
BRIGHAM *The Fast Fourier Transform*
HAMMING *Digital Filters*
OPPENHEIM and SCHAFER *Digital Signal Processing*
OPPENHEIM, et al *Applications of Digital Signal Processing*
RABINER and GOLD *Theory and Applications of Digital Signal Processing*
RABINER and SCHAFER *Digital Processing of Speech Signals*

Advanced Monographs

MCCLELLAN and RADER *Number Theory in Digital Signal Processing*
TRIBOLET *Seismic Applications of Homomorphic Signal Processing*

Seismic Applications of Homomorphic Signal Processing

José Manuel Tribolet

Instituto Superior Técnico
Lisbon, Portugal

Prentice-Hall, Inc., Englewood Cliffs, New Jersey 07632

Library of Congress Cataloging in Publication Data

TRIBOLET, JOSÉ MANUEL. (date)
Seismic applications of homomorphic signal processing.

(Advanced monographs in signal processing)
Bibliography: p.
Includes index.
1. Signal processing. 2. Seismology. 3. Seismic prospecting. I. Title. II. Series.
QE539.T74 551.2′2 78-18385
ISBN 0-13-779801-6

Printed in the United States of America

10 9 8 7 6 5 4 3 2 1

PRENTICE-HALL INTERNATIONAL, INC., *London*
PRENTICE-HALL OF AUSTRALIA PTY. LIMITED, *Sydney*
PRENTICE-HALL OF CANADA, LTD., *Toronto*
PRENTICE-HALL OF INDIA PRIVATE LIMITED, *New Delhi*
PRENTICE-HALL OF JAPAN, INC., *Tokyo*
PRENTICE-HALL OF SOUTHEAST ASIA PTE. LTD., *Singapore*
WHITEHALL BOOKS LIMITED, *Wellington, New Zealand*

To
Magui
and
Bernardo

Contents

Preface

This book is the outcome of three years of doctoral research at M.I.T.'s Department of Electrical Engineering and Computer Science. This research was sponsored by the Portuguese Instituto Nacional de Investigação Científica, to whom I greatfully acknowledge their financial support.

My doctoral program was a rich and enjoyable experience, amidst stimulating and friendly interactions with the members of the Digital Signal Processing Group of M.I.T.'s Research Laboratory of Electronics.

In particular, Professor A. V. Oppenheim has not only had a direct influence on this book, but has also had a profound effect upon the personal and professional growth of the author.

I would also like to express my appreciation to Professor James McClellan, Professor Arthur Baggeroer and Dr. Enders Robinson who have offered many criticisms and corrections that have improved the book. A special thank you to Miss Monica Edelman for her contribution to typing the draft of this book.

But, most of all, I extend my warm thanks to my wife Magui and my son, Bernardo, who contributed their patience and understanding, as well as their personal sacrifice. To them I am particularly indebted.

Instituto Superior Técnico
Lisbon, Portugal

JOSÉ M. TRIBOLET

BIOGRAPHICAL NOTE

José M. Tribolet was born in Tancos, Portugal, on December 20, 1949. He received the Engenheiro Electrotécnico degree from the Instituto Superior Técnico, Lisbon, Portugal, in 1972, and the M.S., E.E., and Sc.D. degrees in electrical engineering and computer science from the Massachusetts Institute of Technology, Cambridge, in 1974, 1975, and 1977 respectively.

From 1970 to 1972 he was an Assistente Eventual of the Instituto Superior Técnico and a Researcher of the Centro de Estudos de Electrónica of the Instituto de Alta Cultura.

From 1972 to 1977 he was a member of the Massachusetts Institute of Technology Research Laboratory of Electronics, with an Instituto Nacional de Investigação Científica Fellowship. During this period his research activities involved the application of homomorphic signal processing to speech and seismic data analysis.

From 1977 to 1978 he was a consultant at the Acoustics Research Department, Bell Laboratories in Murray Hill, N.J., where he worked on adaptive transform coding of speech and on objective quality measures for speech waveform coders.

Mr. Tribolet is now with the Department of Electrical Engineering, Instituto Superior Técnico, Lisbon, Portugal where he is Associate Professor. Mr. Tribolet is a member of Sigma Xi.

I

Introduction

I.1 INTRODUCTION

Seismic signals represent the earth's response to excitations arising from natural phenomena such as earthquakes or from man-made acoustic sources such as those used in exploration geophysics. The purpose of seismic signal processing is to facilitate the interpretation of the received data in terms of the earth's subsurface structure.

The detailed representation of seismic signals involves a relatively complicated model, and the overall processing often requires the use of a set of techniques, each, perhaps, based on different simplifications of the model. One class of seismic analysis techniques is based on a representation of the seismic signal as a convolution of a wavelet representing the seismic wavefront with an impulse train denoting the various wavelet amplitudes and arrival times present in the signal, with the basic signal-processing task being deconvolution.

The purpose of this book is to explore the use of homomorphic signal analysis [1] for seismic deconvolution. The investigation of seismic wavelet estimation by homomorphic filtering will constitute our central topic. Specific strategies for short-time wavelet estimation are developed and illustrated. The resulting wavelet estimates are subsequently used for designing parametric inverse filters which attempt to recover the wavelet's amplitudes and arrival times. The role of homomorphic analysis in parametric inverse filter design for mixed-phase wavelets is discussed.

I.2 SEISMIC DECONVOLUTION

The earth's response as measured by surface receivers is a complex of distinct components, some of them to be considered as signals, since they carry information that enables the potential recovery of the subsurface information, some of them to be regarded as noise, since they either have no relationship with the subsurface structure or are too complex to be considered in a processing scheme. Fortunately, signal and noise waves propagate with different speeds and directions in the earth, thus allowing eventual discrimination by appropriate multichannel processing techniques. Furthermore, noise occupies the low- and high-frequency bands of the seismic spectrum. As a consequence, seismic signals are usually bandpass-filtered prior to sampling and recording. The resulting seismograms, having been stripped of most noise components, essentially consist of many wavelets with different strengths and arrival times, due to the wavefronts that have traveled different source-to-receiver paths and have suffered different degrees of attenuation. This attenuation usually manifests itself in terms of a progressive low-pass filtering effect of the seismic wavelet as it travels farther and farther within the earth. On a short-time basis, however, the wavelet shape is essentially stationary, thus leading to the representation of the seismic trace in terms of a convolution of the wavelet sequence with the corresponding train of arrivals. The goal of the analysis is then to resolve the various arrivals, that is, to deconvolve the impulse train from the composite signal.

A study of the previous application of homomorphic signal processing to seismic deconvolution led to the following three conclusions:

1. The class of homomorphic filters that has previously been used in seismic analysis incorporated a characteristic system that does not permit the processing of bandpass-filtered signals. As a consequence, since the bandpass nature of the seismic data was not taken into account, the out-of-band residual energy, which is dominated by noise, may seriously affect the ability of such homomorphic systems in recovering the signal components.
2. The methods used to implement the characteristic system often led to computational errors due to unreliability of the phase-unwrapping algorithms used to evaluate the complex cepstrum.
3. The impulse-train component of the seismic signal was assumed to be minimum-phase or was made so by exponential weighting. We shall demonstrate that this practice deterministically associates the low-time cepstral information only with the first wavelet arrival. This approach leads typically to impulse-train estimates, where the first arrival is perfectly resolved but the remainder is corrupted by noise. The severity of this problem depends on the particular time-varying nature of the data and of the signal-to-noise ratio.

The facts described above had a major impact in shaping the scope of this book.

I.3 SCOPE OF THE BOOK

The book is divided into two parts. The first part deals with the theory and implementation of homomorphic systems for convolution (Chapters II–V). The second part is dedicated to the development of strategies for seismic data analysis (Chapters VI–XI).

In Chapter II we review the algebraic foundation of homomorphic signal processing, as developed originally by Oppenheim [2]. The canonic representation of homomorphic systems is discussed in detail to provide a framework for the analytical characterization of these systems.

In Chapter III we review the class of homomorphic systems for convolution originally developed by Oppenheim and Schafer [1]. This class of systems is defined by a characteristic system which maps a product of Fourier transforms into a sum of Fourier transforms using a complex logarithmic mapping in the z-domain. It is shown that these systems are not appropriate for the processing of bandpass-filtered signals. To emphasize this fact, these systems are referred to as full-band homomorphic systems.

Given the bandpass characteristics of seismic signals, we develop in Chapter IV a class of homomorphic systems matched to these characteristics. These systems are referred to as bandpass homomorphic systems. Their characteristic system is essentially a logarithmic mapping whose domain has been restricted to encompass only the passband of the input.

In Chapter V we discuss the implementation of homomorphic systems. The principal computational step is the evaluation of the imaginary part of the complex logarithm, usually referred to as phase unwrapping. This problem is thoroughly investigated in this book, leading to an adaptive unwrapping technique that has proved to be highly reliable.

The development of strategies for homomorphic seismic data analysis is preceded by a review of the basic seismic signal models in Chapter VI. Chapter VII is dedicated to a review of the homomorphic deconvolution methods published in the geophysical literature.

In Chapter VIII these methods are assessed in terms of a common analysis framework. This framework is essentially characterized by the representation of seismic traces in terms of time-invariant seismic models and the use of exponential weighting, when necessary, to ensure the minimum-phase character of the impulse train of arrivals. All techniques based on this framework are fundamentally limited by the fact that the low-time values of the trace's complex cepstrum are uniquely determined by the first arrival and depend in no way on subsequent arrivals. As a consequence, this approach is not recommended, in general, for the processing of seismograms, unless the wavelet shape is stationary in time and the signal-to-noise ratio is high.

A more general analysis framework that takes into account the specific time-varying characteristics of seismic traces is developed in Chapter IX. By taking advantage of the slow time variation of the seismic wavelet characteristics, short-time signal models are developed which represent the seismic trace as a series of time-invariant models, each being valid on a particular short-time interval. The short-time window used must not render the impulse train minimum-phase, if the seismic wavelet is to be represented in the low-time cepstral region, independent of the precise positioning of the window.

In Chapter X a strategy for short-time wavelet estimation is proposed which capitalizes on the sensitivity of the cepstral structure of mixed-phase aperiodic impulse trains to time-domain amplitude perturbations. The procedure consists of using different short-time windows on the same seismic segment. Each of these windowed segments is homomorphically filtered, using a cepstral gate around the time origin. The seismic wavelet is then estimated using appropriate averaging procedures, either in the time domain or in the cepstral domain.

Finally, in Chapter XI parametric deconvolution strategies for the recovery of the reflector series are outlined, based on the combination of homomorphic wavelet estimation with optimum-lag Wiener spiking filtering and with homomorphic predictive inverse filtering.

The methods proposed in this book have been tested using synthetic data models. We have sought to ensure that the synthetic seismograms employed as test vehicles are realistic, based on a reasonable and complete earth model. In particular, both frequency-dependent attenuation and additive noise have been incorporated in the data models, leading to seismograms that deviate in a well-defined way from idealized convolutional models. Using such synthetic seismograms, a qualitative evaluation of the algorithms proposed for seismic analysis can be done in terms of the underlying earth structure.

The ultimate test of these new algorithms will, of course, be their performance in the face of actual data. This is an area for future research.

II

The Algebraic Theory of Homomorphic Systems

II.1 INTRODUCTION

When we are faced with the problem of filtering signals that have been added, we very often use a linear filter. A principal reason is that linear filters are analytically convenient for dealing with signals that have been added, since, as a consequence of the principle of superposition, the behavior of a linear system L for the sum of signals is the sum of the responses, that is,

$$L[ax_1[n] + x_2[n]] = aL[x_1[n]] + L[x_2[n]] \tag{2-1}$$

In many physical situations, however, we encounter signals that may be represented in terms of components that are combined according to a rule other than addition (e.g., convolution or multiplication). In such cases it might be equally advantageous to filter the signals through systems that are matched to this rule in the same way that linear systems are matched to addition.

This observation leads to consideration of classes of nonlinear systems that obey a generalized principle of superposition. An approach to characterizing such nonlinear systems was presented by Oppenheim in 1965 [2]. In this approach, signals are regarded as vectors in vector spaces. This representation is, in fact, a very convenient way of formalizing the basic rules by which signals are combined with other signals or with scalars. A typical case is the signal vector space, where vector addition equals signal addition and scalar multiplication equals the multiplication of a signal by a number. This particular vector space is a Hilbert space under the usual inner product

$$\sum_{n=-\infty}^{\infty} x[n]y^{\dagger}[n]$$

where † denotes complex conjugate.

Signals may be represented as vectors in vector spaces using a variety of definitions of vector addition and scalar multiplication. A vector space may be defined, for example, where vector addition is the product of two signals and scalar multiplication is exponentiation. Alternatively, we may define a vector space using convolution as vector addition and z-transform exponentiation as scalar multiplication. Classes of nonlinear systems may then be defined in terms of algebraically linear transformations between signal vector spaces. To emphasize this fact, such nonlinear systems have been termed *homomorphic systems*.

In this book we are concerned with the analysis of signals that are combined by convolution, with the basic processing task being deconvolution. The analytical study of the class of homomorphic systems for convolution will be conducted in Chapters III, IV, and V. We precede such study by reviewing in this chapter the algebraic theory of homomorphic systems with the purpose of establishing the notation and framework for the remainder of this study. In particular, we reinterpret the fundamental results on the canonical representation of homomorphic systems and show that, if the input space to a homomorphic system is not a Hilbert space, then one must be able to find a homomorphic mapping between the input vector space into a subspace that is a Hilbert space. It is within this framework that we shall later interpret the common practice of input normalization in homomorphic signal analysis.

The results presented in this chapter draw heavily on the concepts and theorems of the linear algebra of vector spaces. The reader is referred to [2] or to any standard linear algebra textbook (e.g., [3]) for a formal definition of these concepts.

II.2 GENERALIZED PRINCIPLE OF SUPERPOSITION

Let H be a transformation between signal vector spaces and let us denote by $\square$ and by $:$ the rules of vector addition and scalar multiplication in the input vector space (IVS) for H. Similarly, let us denote by $\bigcirc$ and $\rceil\!\lfloor$ the rules for vector addition and scalar multiplication in the output vector space (OVS) of H.

The system H is said to satisfy a *generalized principle of superposition* if

$$H[(C: x_1[n]) \square (x_2[n])] = (C \rceil\!\lfloor H[x_1[n]]) \bigcirc (H[x_2[n]]) \qquad (2\text{-}2)$$

Nonlinear systems of this type have been called *homomorphic systems,* to emphasize the fact that they are represented by *algebraically linear transformations* from the IVS onto the OVS. Such systems are represented as in Figure 2.1.

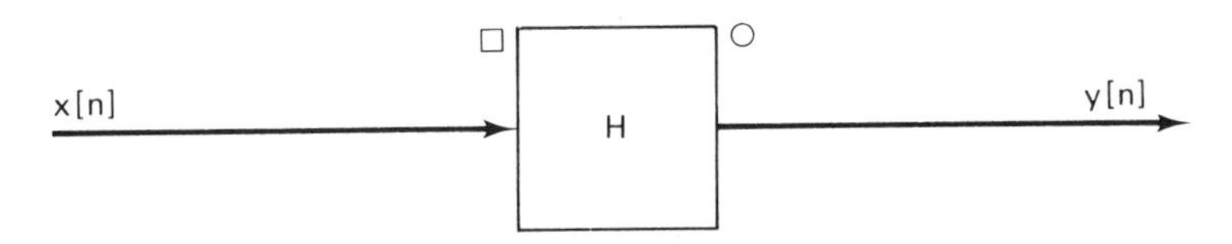

Figure 2.1 Representation of a homomorphic system

II.3 CANONICAL DECOMPOSITION OF HOMOMORPHIC SYSTEMS

A central result in homomorphic systems theory was Oppenheim's derivation of a canonical representation for homomorphic systems [2]. Because of its importance, we shall review below the framework within which this result is valid. The key to the derivation of this result lies in the fact that if continuity is imposed on an algebraically linear transformation, then the range of the transfor-

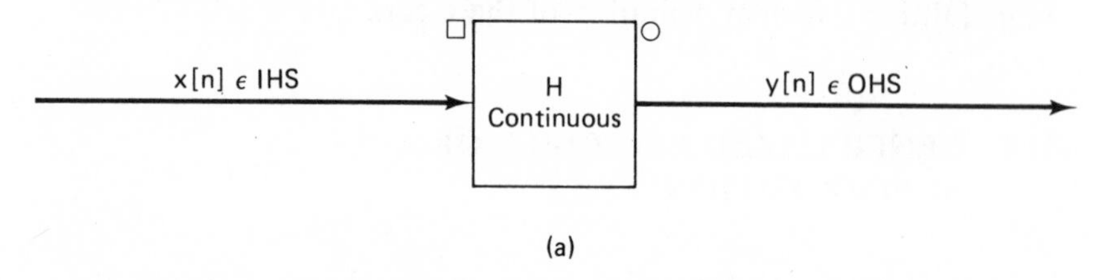

(a)

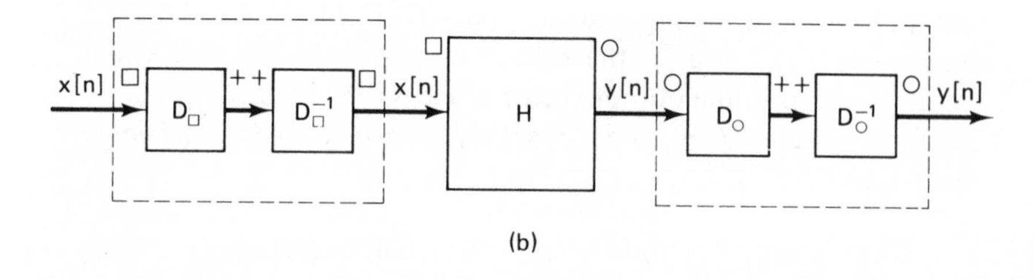

(b)

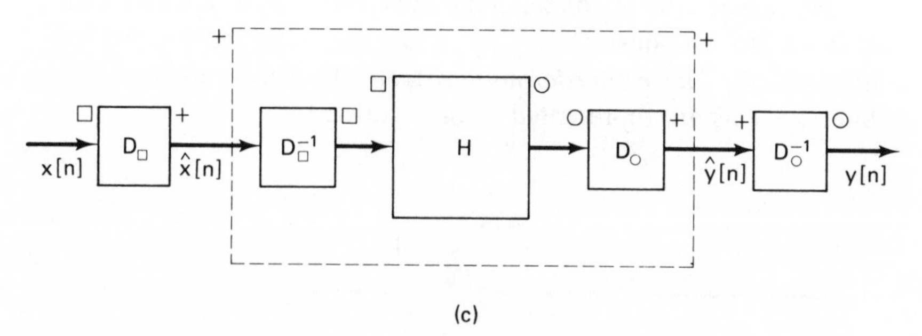

(c)

Figure 2.2 Framework for the canonical decomposition of homomorphic systems

mation will be a separable Hilbert space, whenever the domain is a separable Hilbert space.

Consider, then, a continuous homomorphic transformation H, as in Figure 2.2a, and assume its IVS to be a Hilbert space. In that case it is always possible to define an isomorphism $D_{\Box}$ between the input Hilbert space (IHS) and the standard additive Hilbert space. Similarly, one can always define an isomorphism $D_{\bigcirc}$ between the output Hilbert space (OHS) and the additive Hilbert space.

Denoting by $D_{\Box}^{-1}$ and $D_{\bigcirc}^{-1}$ the inverse of $D_{\Box}$ and $D_{\bigcirc}$, we will not affect the transformation H if we cascade at its input the system $D_{\Box}$ followed by $D_{\Box}^{-1}$, and at its output the system $D_{\bigcirc}$ followed by $D_{\bigcirc}^{-1}$, as illustrated in Figure 2.2b. Noting that the cascade of $D_{\Box}^{-1}$ with H and $D_{\bigcirc}$ is a conventional linear system as depicted in Figure 2.2c, and denoting such linear systems as L, we obtain the conventional decomposition illustrated in Figure 2.3.

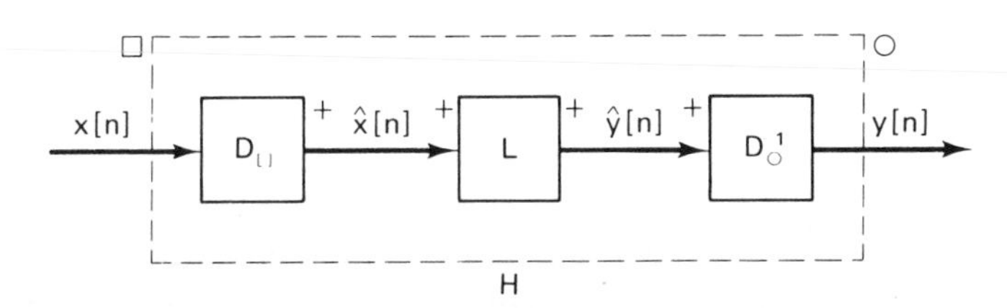

Figure 2.3 Canonical decomposition

It is clear from Figure 2.2c that all the filtering ability of a homomorphic system, represented in the canonical form, lies in the choice of its linear system component L. Since the systems $D_{\Box}$ and $D_{\bigcirc}$ depend only on the characterization of the IHS and the OHS of the system H, they have been termed *characteristic systems* for the operations $\Box$ and $\bigcirc$, respectively.

The characteristic system $D_{\Box}$ is then an invertible continuous homomorphic system, which transforms the combination of signals $x_1[n]$ and $x_2[n]$ according to the rule $\Box$ into a conventional linear combination of corresponding signals $\hat{x}_1[n]$ and $\hat{x}_2[n]$, that is,

$$\begin{aligned}\hat{x}[n] = D_\square[x_1[n] \,\square\, x_2[n]] &= D_\square[x_1[n]] + D_\square[x_2[n]] \\ &= \hat{x}_1[n] + \hat{x}_2[n]\end{aligned} \tag{2-3}$$

The system L, being linear, is such that

$$L[\hat{x}_1[n] + \hat{x}_2[n]] = L[\hat{x}_1[n]] + L[\hat{x}_2[n]] = \hat{y}_1[n] + \hat{y}_2[n] \tag{2-4}$$

Finally, the system $D_\bigcirc^{-1}$ transforms addition into the operation $\bigcirc$, so that

$$\begin{aligned}y[n] = D_\bigcirc^{-1}[\hat{y}_1[n] + \hat{y}_2[n]] &= D_\bigcirc^{-1}[\hat{y}_1[n]] \bigcirc D_\bigcirc^{-1}[\hat{y}_2[n]] \\ &= y_1[n] \bigcirc y_2[n]\end{aligned} \tag{2-5}$$

In summary, any homomorphic system whose IVS is a Hilbert space can be decomposed as in the canonic representation of Figure 2.3. Such systems will be henceforth referred to as *canonical homomorphic systems.*

Noncanonical homomorphic systems, that is, homomorphic systems with inputs from an IVS that is not necessarily a Hilbert space, can be decomposed as a cascade of a noncanonical system N with a canonical system as depicted in Figure 2.4, where the system N maps the IVS onto a Hilbert subspace of the IVS. It can be shown [2] that:

1. All signals $s[n]$ in the IVS, defined as

$$N[s[n]] = N[e[n]]$$

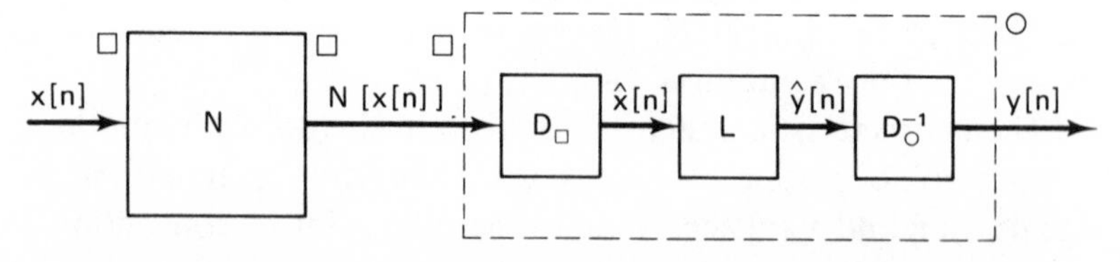

Figure 2.4 General structure for noncanonical homomorphic systems

where $e[n]$ is the identity element in IVS, must form a subspace S_N of IVS.

2. Consider the quotient space IVS/S_N. Let $x_1[n] \square S_N$ be any coset in the quotient space. If $x_2[n]$ is a signal in $x_1[n] \square S_N$, then

$$N[x_1[n]] = N[x_2[n]]$$

That is, any given coset in the quotient space IVS/S_N must be associated with a unique output of N, and each coset must produce a different output. It follows, then, that the quotient space IVS/S_N is isomorphic to the Hilbert space at the input of $D_\square$.

In practice, the key to the synthesis of a homomorphic system matched at the input to the operation $\square$ is the choice of the characteristic system $D_\square$. By selecting a function that maps $\square$ to addition and by analytically enforcing the continuity and invertibility of this mapping, one is equivalently determining the Hilbert space of IVS, which must constitute the range of N. All that remains to be done, then, is the careful characterization of the mapping N and the verification that such mapping has the properties stated above.

The class of homomorphic systems to be considered in this book are systems matched to convolution at both input and output. The characteristic systems used are essentially characterized by a logarithmic operation in the z-transform domain. The mappings N correspond to very simple normalization procedures by which a signal is appropriately time-shifted or is sign-reversed prior to being transformed by the characteristic system. We shall henceforth refer to N as the *input normalization mapping.*

II.4 HOMOMORPHIC SIGNAL ANALYSIS

Once the characteristic system $D_\square$ and the input normalization mapping N have been determined, all the filtering versatility of a homomorphic system lies in its linear system component. For example, if we wish to recover $x_1[n]$ from the signal

$$x[n] = x_1[n] \,\square\, x_2[n]$$

we must choose the linear system so that its output $\hat{y}[n]$ is

$$\hat{y}[n] = L[\hat{x}[n]] = \hat{x}_1[n]$$

Then, with $D_{\bigcirc} = D_{\square}$,

$$y[n] = D_{\square}^{-1}[\hat{x}_1[n]] = N[x_1[n]]$$

Thus, to perfectly separate the normalized signal components at the output of a homomorphic system, we must be able to perfectly separate $\hat{x}_1[n]$ and $\hat{x}_2[n]$ using a linear filter. How well we can approach this ideal situation depends on the operation $\square$ and the properties of the signals $x_1[n]$ and $x_2[n]$.

II.5 SUMMARY

Homomorphic systems are nonlinear systems that satisfy the generalized principle of superposition. An important class of homomorphic systems are canonical homomorphic systems. Such systems may be classified according to their input and output characteristic systems. Characteristic systems may often be analytically designed, thus avoiding the need for an algebraic approach to the synthesis of homomorphic systems. Canonical homomorphic systems may also be used in the synthesis of more general classes of homomorphic systems, when preceded by appropriate noncanonical normalization mappings.

III

Homomorphic Systems for Convolution: Full-band Systems

III.1 INTRODUCTION

An important class of seismic analysis techniques is based on a representation of the seismic signal as a convolution of components with the basic signal processing task being *deconvolution*.

There are a number of approaches typically available for carrying out a deconvolution. One of the most common is the use of linear inverse filtering, that is, processing the composite signal through a linear filter whose frequency response is the reciprocal of the Fourier transform of one of the signal components. Obviously, in order to use inverse filtering, such components must be known or estimated.

Homomorphic signal processing represents an alternative approach to seismic deconvolution. The investigation of this alternative constitutes the central topic of this book. In the next three chapters we shall take a first step toward this goal, by conducting an in-depth study of the class of homomorphic systems for convolution.

III.2 HOMOMORPHIC SYSTEMS FOR CONVOLUTION

Any nonlinear system H that satisfies the convolutional superposition property illustrated in Figure 3.1 is called a *homomorphic*

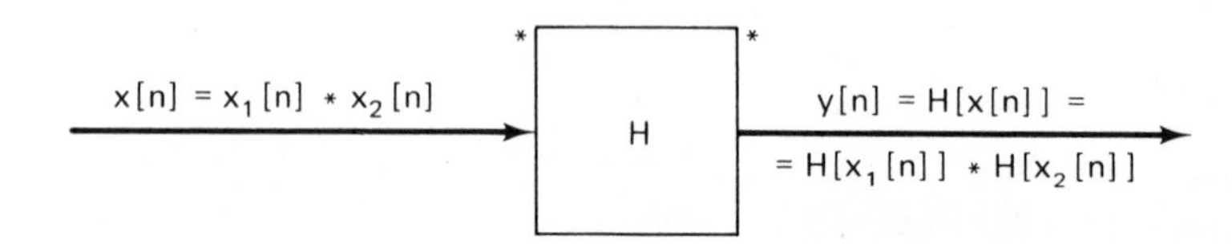

Figure 3.1 Convolutional superposition property

system for convolution. The first applications of homomorphic systems for convolution were made by Oppenheim and Schafer [4] in the areas of speech analysis and echo removal and detection. Today, homomorphic convolutional systems are applied to a wide variety of problems, ranging from speech vocoding [5] and systems identification [6] to EEG analysis [7] and seismic dereverberation [8]. The class of canonical homomorphic systems for convolution satisfies the canonical representation of Figure 3.2.

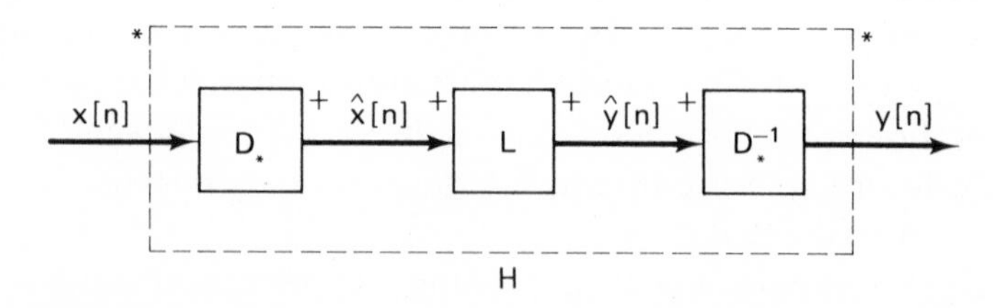

Figure 3.2 Canonical representation of homomorphic systems for convolution

The present chapter is dedicated to the study of a class of homomorphic systems for convolution, defined by a characteristic system $D_\star$, such that the z-transform of its output is equal to the complex logarithm of its input; that is, if

$$\hat{x}[n] = D_\star[x[n]] \tag{3-1a}$$

then

$$\hat{X}(z) = \log X(z) \tag{3-1b}$$

This class of systems was first proposed by Oppenheim and Schafer [4] in their original work on homomorphic filtering and has provided the basis for all subsequent applications of homomorphic systems. However, as discussed in Chapter II, the choice of a characteristic system is equivalent to the choice of a particular input space for the corresponding class of systems. It is therefore necessary to investigate whether or not the characteristics of such a space are compatible with the characteristics of the signals one wishes to analyze.

III.3 THE CHARACTERISTIC SYSTEM

The key to the mathematical representation of the characteristic system $D_\star$, as in equations (3-1), is based on the fact that the z-transform of a convolution of two sequences is equal to the product of the z-transforms of each of the sequences. That is, letting

$$x[n] = x_1[n] \star x_2[n] \tag{3-2a}$$

it follows that

$$X(z) = X_1(z) \cdot X_2(z) \tag{3-2b}$$

Thus, the z-transform operation can be viewed as a homomorphic transformation with convolution as the input operation and multiplication as the output operation.

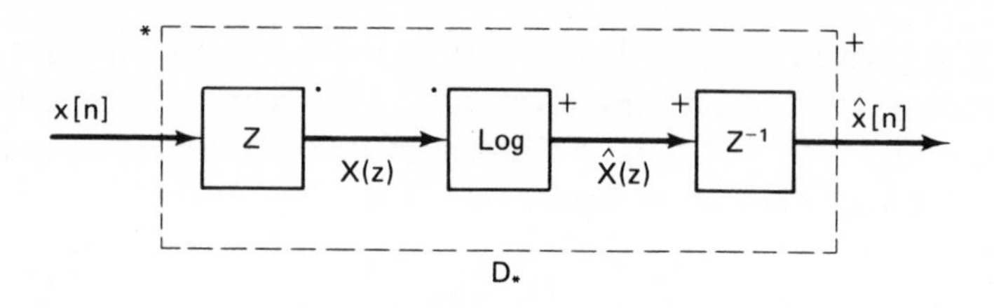

Figure 3.3 Representation of the complex logarithmic characteristic system $D_\star$

On the other hand, z-transforms are linear transformations in the conventional sense. Thus, we may represent the characteristic system $D_\star$ in terms of a cascade of three homomorphic systems with compatible interface operations, as illustrated in Figure 3.3. Note that the inverse z-transform is not essential in the representation of the characteristic system, since at its input, signals are already additively combined. The advantage of z^{-1} is, however, that it allows the outputs of $D_\star$ to be sequences. We shall henceforth refer to the characteristic system of Figure 3.3 as the complex logarithmic characteristic system.

Implicit in the definition of equation (3-1) is the fact that both z-transforms $X(z)$ and $\hat{X}(z)$ converge in some annular region in the z-plane.

We shall require the region of convergence of $\hat{X}(z)$ to include the unit circle, thus restricting the sequences $\hat{x}[n]$ to be stable, which is certainly a sound practical requirement. This restriction is also theoretically sound, since the additive vector space formed by all stable sequences is a Hilbert space, as required by the canonical representation theorem. The most important implication of the analyticity of $\hat{X}(z)$ on the unit circle is that $\hat{X}(e^{j\omega})$ must be a continuous function of ω. Letting

$$
\begin{aligned}
X(e^{j\omega}) &= X_R(e^{j\omega}) + jX_I(e^{j\omega}) \\
&= |X(e^{j\omega})|\, e^{j \arg|X(e^{j\omega})|}
\end{aligned}
\tag{3-3a}
$$

it follows that

$$\begin{aligned}\hat{X}(e^{j\omega}) &= \hat{X}_R(e^{j\omega}) + j\hat{X}_I(e^{j\omega}) \\ &= \log|X(e^{j\omega})| + j\arg|X(e^{j\omega})|\end{aligned} \tag{3-3b}$$

Thus, both $\log|X(e^{j\omega})|$ and $\arg|X(e^{j\omega})|$ must be continuous functions of ω.

Continuity of log $|X(e^{j\omega})|$

The log magnitude function is continuous on the unit circle if $X(z)$ and its inverse $X^{-1}(z)$ are both analytic on the unit circle. Thus, the inputs $x[n]$ are restricted to the class of *stable* sequences with *stable inverses*. We shall henceforth refer to sequences within this class as *SSI sequences*. In particular, low-pass, high-pass, and bandpass signals are not allowable as inputs to this class of homomorphic systems. To emphasize such a fact, these systems shall be referred to as *full-band homomorphic systems*.

Continuity of arg $|X(e^{j\omega})|$

Let us now investigate the continuity of the phase $\arg|X(e^{j\omega})|$. Since the phase of a complex number is a multivalued function, the continuity of $\arg|X(e^{j\omega})|$ depends not only on specific properties of $x[n]$ but also on the definition of the complex logarithm. One approach to such a definition is to capitalize on the analyticity of $\hat{X}(z)$ within its region of convergence and to obtain the complex logarithm by integration of its derivative. If we assume a single-valued differentiable complex logarithm, then

$$\frac{d\hat{X}(z)}{dz} = \frac{d\log[X(z)]}{dz} = \frac{1}{X(z)}\frac{dX(z)}{dz} \tag{3-4}$$

Evaluating this logarithmic derivative on the unit circle, we obtain

$$\hat{X}'(e^{j\omega}) = \hat{X}'_R(e^{j\omega}) + j\hat{X}'_I(e^{j\omega}) = \frac{X'(e^{j\omega})}{X(e^{j\omega})} \tag{3-5}$$

where the prime denotes differentiation with respect to ω. Thus, the phase derivative satisfies

$$\frac{d \arg|X(e^{j\omega})|}{d\omega} = \hat{X}'_I(e^{j\omega}) = \frac{X_R(e^{j\omega})X'_I(e^{j\omega}) - X_I(e^{j\omega})X'_R(e^{j\omega})}{|X(e^{j\omega})|^2} \quad (3\text{-}6)$$

It can be shown that the phase derivative is bounded on the unit circle whenever $x[n]$ is a SSI sequence. The phase derivative is an even function of ω whenever $x[n]$ is a real sequence. Furthermore, it can be shown that

$$\frac{d \arg|X_1(e^{j\omega})X_2(e^{j\omega})|}{d\omega} = \frac{d \arg|X_1(e^{j\omega})|}{d\omega} + \frac{d \arg|X_2(e^{j\omega})|}{d\omega} \quad (3\text{-}7)$$

The definition of the phase function by integration of its derivative is unique, to within an integration constant. This constant must be determined, according to the requirement that the logarithm be a homomorphic mapping between multiplication and addition, as represented in Figure 3.3. Thus, one must have

$$\arg|X_1(e^{j\omega})X_2(e^{j\omega})| = \arg|X_1(e^{j\omega})| + \arg|X_2(e^{j\omega})| \quad (3\text{-}8)$$

which, for real $x[n]$, can only be satisfied if the integration constant is chosen so that

$$\arg|X(e^{j\omega})|_{\omega=0} = 0 \quad (3\text{-}9a)$$

Since

$$X(e^{j\omega})|_{\omega=0} = \sum_{n=-\infty}^{+\infty} x[n] \quad (3\text{-}9b)$$

it follows that only signals with a positive dc component will be compatible with such requirement. Finally, $\arg|X(e^{j\omega})|$ must be, of course, periodic in ω, with period 2π. Being odd, owing to equation (3-9a), and continuous, the phase must satisfy

$$\arg|X(e^{j\omega})|_{\omega=\pi} = 0 \quad (3\text{-}10a)$$

Since

$$\arg|X(e^{j\omega})|_{\omega=\pi} = \int_0^{\pi} \left[\frac{d\arg|X(e^{j\omega})|}{d\omega}\right] d\omega \tag{3-10b}$$

and the phase derivative is an even function of ω, we conclude that only signals with zero mean phase derivative are compatible with the requirement above.

It can be shown that the set of all SSI sequences with positive dc component and zero mean phase derivative form a vector space with convolution as vector addition. Thus, since the complex logarithmic transformation defined above is a continuous invertible mapping from this IVS onto the Hilbert space formed by the vectors $\hat{x}[n]$ under addition, we conclude that such mapping is a characteristic system and that the corresponding IVS is a Hilbert space.

The inverse of the complex logarithmic characteristic system, $D_\star^{-1}$, is depicted in Figure 3.4.

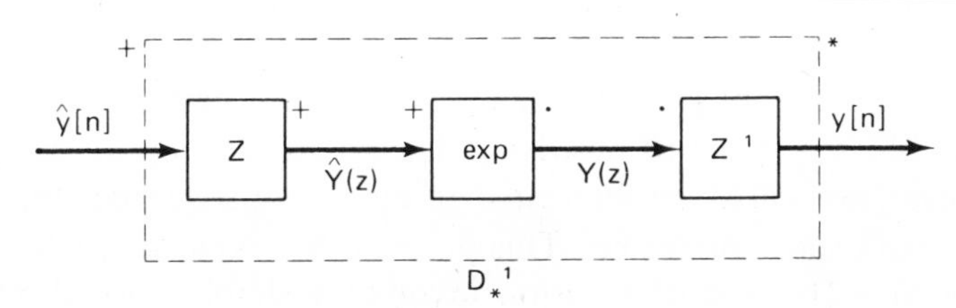

Figure 3.4 Representation of the inverse characteristic system $D_\star^{-1}$

III.4 INPUT NORMALIZATION

We have just shown that the input space to the characteristic system $D_\star$ is a Hilbert space formed by all SSI sequences with positive dc component and zero mean phase derivative. To see how restrictive these attributes really are, let us consider a real full-band sequence $x[n]$ with a rational z-transform $X(z)$. It can be shown that the most general rational z-transform of such sequence $x[n]$ can be written

$$X(z) = s_x A_x z^{r_x} \frac{\prod_{k=1}^{m_i} (1 - a_k z^{-1}) \prod_{k=1}^{m_0} (1 - b_k z^{+1})}{\prod_{k=1}^{p_i} (1 - c_k z^{-1}) \prod_{k=1}^{p_0} (1 - d_k z^{+1})} \tag{3-11}$$

where s_x equals $+1$ or -1, A_x is a positive real value, r_x is an integer, and $|a_k|$, $|b_k|$, $|c_k|$, and $|d_k|$ are all less than unity.

It can also be shown that

$$s_x = \text{SIGN}\left(\sum_{n=-\infty}^{+\infty} x[n]\right) \tag{3-12a}$$

where SIGN denotes the signum function, and that

$$r_x = \frac{1}{2\pi} \int_{-\pi}^{\pi} \left[\frac{d \arg |X(e^{j\omega})|}{d\omega}\right] d\omega \tag{3-12b}$$

We shall henceforth refer to s_x as the *polarity* of $x[n]$ and to r_x as the *lag* of $x[n]$.

Thus, the input space to the characteristic system $D_\star$ is formed by all SSI sequences of positive polarity and zero lag. In general, however, we might be interested in analyzing sequences that do not satisfy such conditions. This problem has been recognized for some time [1], and its solution involves a simple normalization procedure, to be described next.

Let S be the set of all SSI sequences with arbitrary polarities and lags and with rational z-transforms, and let N be a transformation defined as

$$N[x[n]] = s_x x[n + r_x] \tag{3-13}$$

In other words, the normalization N corresponds to a *sign reversal* for sequences with negative polarity and to a *time shift* by an amount equal to the signal lag. Thus, since any sequence $N[x[n]]$ has positive polarity and zero lag, the image of S under N is now a subset of the IHS to $D_\star$.

For the normalization N to be well defined, from a theoretical point of view, one needs to extend its domain to an IVS that contains S as a subset, since in fact S is not a vector space under convolution. It can be shown that such a vector space is formed by all SSI sequences $x[n]$ with z-transforms of the form

$$X(z) = s_x A_x z^{r_x} \frac{\prod_{k=1}^{m_i} (1 - a_k z^{-1})^{\alpha_k} \prod_{k=1}^{m_0} (1 - b_k z^{+1})^{\beta_k}}{\prod_{k=1}^{p_i} (1 - c_k z^{-1})^{\gamma_k} \prod_{k=1}^{p_0} (1 - d_k z^{+1})^{\eta_k}} \tag{3-14}$$

where s_x, A_x, $|a_k|$, $|b_k|$, $|c_k|$, and $|d_k|$ are as in equation (3-11) and r_x, α_k, β_k, γ_k, and η_k are now, in general, rational numbers. Since any rational number can be written as the ratio of two integers, the class of z-transforms defined above has been referred to as fractional pole–zero irrational models [6].

The analytical representation of N for this general class of sequences can no longer be defined in terms of time shifts as in equation (3-13). The correct representation keeps, however, a simple analytical form in the z-transform domain:

$$X(z) \xrightarrow{N} s_x z^{-r_x} X(z) \tag{3-15}$$

The normalization mapping N is theoretically correct since it fully satisfies the conditions required for it to be cascaded with a canonic homomorphic system for convolution, leading to the synthesis of a general homomorphic system for convolution with input space the IVS defined above. In fact, it can be shown that:

1. N is a homomorphic system for convolution, since given

$$x[n] = x_1[n] \star x_2[n]$$

it follows that

$$s_x = s_{x_1} \cdot s_{x_2}$$

and

$$r_x = r_{x_1} + r_{x_2}$$

and thus

$$N[x[n]] = N[s_x x[n + r_x]] = N[x_1[n]] \star N[x_2[n]]$$

2. The set S_N of elements with z-transform of the form $s_x z^{+r_x}$, $s_x = \pm 1$, r_x any rational, forms a subspace of IVS. Furthermore, each member $s[n]$ of S_N satisfies

$$N[s[n]] = \delta[n] = D_\star^{-1}[\hat{\delta}[n]]$$

where

$$\hat{\delta}[n] = 0, \qquad \forall\, n$$

3. Consider the coset $x[n] \star S_N$, in the quotient space IVS/S_N. Thus, all elements within $x[n] \star S_N$ differ only with respect to the corresponding values of s_x and r_x. Therefore, all elements within this coset will be mapped under N into a unique element $N[x[n]]$. Thus, the quotient space IVS/S_N is isomorphic to the input Hilbert space of $D_\star$.

From a practical point of view, and in the context of deconvolution by homomorphic signal processing, the input normalization amounts to the normalization of the signal components. Thus, their true polarities and lags cannot be determined by this technique. This is analogous in linear systems theory to the situation in which two signals, each with nonzero dc levels, have been added. Using linear filtering on the sum of the signals, the true dc levels of the components can never be determined.

For most deconvolution problems, however, the loss of polarity and lag information does not represent a serious problem, since such information can often be measured by other means.

III.5 THE COMPLEX CEPSTRUM

Now that the characteristic system $D_\star$ has been defined and precise conditions at its input have been determined, let us concentrate on its output.

The output $\hat{x}[n]$ of the system $D_\star$ is referred to as the *complex cepstrum* of the input $x[n]$. This terminology is motivated by the relationship between this transformation and the cepstrum as proposed by Bogert, Healy, and Tukey [9] for the detection of echoes. Specifically, the cepstrum $c[n]$ of a signal $x[n]$ was defined as the power spectrum of the logarithm of the power spectrum of that signal. Since the cepstrum was directed toward echo detection rather than deconvolution, retention of phase information was not important. Thus, it does not utilize phase information and involves the logarithm of real, positive values.

By contrast, the output of the system $D_\star$ is referred to as the complex cepstrum, in reference to the fact that it requires the use of the complex logarithm. It is important to note, however, that the complex cepstrum of a real sequence is a *real-valued sequence.*

Defining the even and odd components of the complex cepstrum $\hat{x}_e[n]$ and $\hat{x}_0[n]$ as

$$\hat{x}[n] = \hat{x}_e[n] + \hat{x}_0[n] \tag{3-16a}$$

where

$$\hat{x}_e[n] = \frac{\hat{x}[n] + \hat{x}[-n]}{2} = \hat{x}_e[-n] \tag{3-16b}$$

and

$$\hat{x}_0[n] = \frac{\hat{x}[n] - \hat{x}[-n]}{2} = -\hat{x}_0[-n] \tag{3-16c}$$

it follows that

$$\hat{x}_e[n] = \tfrac{1}{2}\,\mathrm{IFT}\{\log |X(e^{j\omega})|^2\} \tag{3-17a}$$

and

$$\hat{x}_0[n] = \mathrm{IFT}\{\arg |X(e^{j\omega})|\} \tag{3-17b}$$

Thus, the cepstrum $c[n]$ as defined by Bogert, Healy, and Tukey is proportional to the even part of the complex cepstrum.

The sequence $\hat{x}_e[n]$ is also referred to as the *real cepstrum*. The sequences $\hat{x}_e^2[n]$ and $\hat{x}_0^2[n]$ have been referred to in the literature [10] as the *power cepstrum* and *phase cepstrum*, respectively.

III.6 PROPERTIES OF THE COMPLEX CEPSTRUM

In carrying out a deconvolution using homomorphic filtering, an appropriate choice for the linear system L must be made. Given the fact that the complex cepstrum of a convolution of two (or more) signals is the sum of their individual complex cepstra, the choice of the linear system is intimately connected to the specific characteristics of each of the signal components in the cepstral domain.

In order to investigate these characteristics, we shall first restrict our attention to sequences $x[n]$, with a rational z-transform as in equation (3-11), with m_i zeros and p_i poles inside the unit circle and m_0 zeros and p_0 poles outside it. The sequence $x[n]$ is thus, in general, *mixed-phase* [1].

It is useful to associate all the poles and zeros inside the unit circle in a single factor, defining a normalized *minimum-phase* sequence $x_{\min}[n]$, whose z-transform is

$$X_{\min}(z) = \frac{\prod_{k=1}^{m_i} (1 - a_k z^{-1})}{\prod_{k=1}^{p_i} (1 - c_k z^{-1})} \tag{3-18a}$$

Similarly, we define a normalized *maximum-phase* sequence $x_{\max}[n]$, whose z-transform is

$$X_{\max}(z) = \frac{\prod_{k=1}^{m_0} (1 - b_k z^{+1})}{\prod_{k=1}^{p_0} (1 - d_k z^{+1})} \tag{3-18b}$$

Thus, the sequence $x[n]$ satisfies

$$x[n + r_x] = s_x A_x [x_{\min}[n] \star x_{\max}[n]] \tag{3-19}$$

where $x_{\min}[n] = 0, n < 0$, and $x_{\max}[n] = 0, n > 0$.

The normalized sequence $N[x[n]]$ is then given by

$$N[x[n]] = A_x[x_{\min}[n] \star x_{\max}[n]] \tag{3-20a}$$

and it follows that

$$\hat{X}(z) = \log A_x + \log X_{\min}(z) + \log X_{\max}(z) \tag{3-20b}$$

Consequently,

$$\hat{x}[n] = \log A_x \delta[n] + \hat{x}_{\min}[n] + \hat{x}_{\max}[n] \tag{3-20c}$$

Using the well-known power-series expansions,

$$\log(1 - az^{-1}) = -\sum_{n=1}^{\infty} \frac{a^n}{n} z^{-n}, \qquad |z| > |a| \tag{3-21a}$$

and

$$\log(1 - az^{+1}) = -\sum_{n=1}^{\infty} \frac{a^n}{n} z^{+n}, \qquad |z| < |a^{-1}| \tag{3-21b}$$

it follows that

$$\hat{x}_{\min}[n] = \left[-\sum_{k=1}^{m_i} \frac{a_k^n}{n} + \sum_{k=1}^{p_i} \frac{c_k^n}{n}\right] u[n-1] \tag{3-22a}$$

and

$$\hat{x}_{\max}[n] = \left[-\sum_{k=1}^{m_0} \frac{b_k^n}{n} + \sum_{k=1}^{p_0} \frac{d_k^n}{n}\right] u[-n-1] \tag{3-22b}$$

From the equations above, we observe the following properties of the complex cepstrum.

PROPERTY 1

The complex cepstrum decays at least as fast as $1/n$. Specifically,

$$|\hat{x}[n]| < C\left|\frac{\alpha^n}{n}\right|, \qquad -\infty < n < \infty$$

where C is a constant and α equals the maximum of $|a_k|$, $|b_k|$, $|c_k|$, and $|d_k|$.

PROPERTY 2

If $x[n]$ is of finite duration ($m_o = m_i = 0$), $\hat{x}[n]$ will nevertheless have infinite duration.

PROPERTY 3

The complex cepstrum of a minimum-phase sequence is zero for $n < 0$. The complex cepstrum of a maximum-phase sequence is zero for $n > 0$.

It can be shown that these properties are also valid for the class of sequences with fractional pole–zero irrational z-transforms, as in equation (3-14). The complex cepstrum was derived above in terms of the poles and zeros of $X(z)$. It is also possible to obtain an implicit expression, directly relating $\hat{x}[n]$ with $x[n]$, which in certain cases reduces to a very useful recursion formula.

This implicit relation may be derived as follows. Assuming $x[n]$ to be normalized, then

$$-z\frac{d}{dz}\hat{X}(z) = \frac{1}{X(z)} \cdot \left[-z\frac{d}{dz}X(z)\right] \tag{3-23a}$$

and it follows that

$$n\hat{x}[n] \star x[n] = nx[n] \tag{3-23b}$$

Since for (normalized) mixed-phase signals, both $x[n]$ and $\hat{x}[n]$ are two-sided sequences, the implicit equation above cannot be rearranged into a recursion formula for the computation of $\hat{x}[n]$ from $x[n]$, or vice versa.

However, both $x_{\min}[n]$ and $\hat{x}_{\min}[n]$ are causal sequences. Similarly, both $x_{\max}[n]$ and $\hat{x}_{\max}[n]$ are anticausal sequences. One may then derive from equation (3-23b) recursive formulas which are valid for the class of minimum-phase and maximum-phase sequences. We can show the following two properties.

PROPERTY 4

The complex cepstrum of a minimum-phase sequence satisfies

$$\hat{x}_{\min}[n] = x_{\min}[n] - \sum_{k=1}^{n-1}\left(\frac{k}{n}\right)\hat{x}_{\min}[k]x_{\min}[n-k] \qquad \text{(3-24a)}$$

Similarly,

$$x_{\min}[n] = \sum_{k=1}^{n}\left(\frac{k}{n}\right)\hat{x}_{\min}[k]x_{\min}[n-k] \qquad \text{(3-24b)}$$

PROPERTY 5

The complex cepstrum of a maximum-phase sequence satisfies:

$$\hat{x}_{\max}[n] = x_{\max}[n] - \sum_{k=n+1}^{-1}\left(\frac{k}{n}\right)\hat{x}_{\max}[k]x_{\max}[n-k] \qquad \text{(3-25a)}$$

Similarly,

$$x_{\max}[n] = \sum_{k=n}^{-1}\left(\frac{k}{n}\right)\hat{x}_{\max}[k]x_{\max}[n-k] \qquad \text{(3-25b)}$$

These recursive relationships will be called to play an important role later in this book. In general, they can be used as a realization of $D_\star$, whenever the inputs are known to be minimum-phase or maximum-phase. They can always be used as a realization of $y[n] = D_\star^{-1}[\hat{y}[n]]$, by independently computing the minimum- and maximum-phase signal components $y_{\min}[n]$ and $y_{\max}[n]$ and then convolving the results according to the equation

$$y[n] = e^{\hat{y}[0]}[y_{\min}[n] \star y_{\max}[n]] \qquad \text{(3-26)}$$

The properties of the complex cepstrum derived above are valid in general. Now, we shall investigate a number of properties regarding the cepstral structure of those classes of signals often encountered in deconvolution problems.

Many physical phenomena are characterized by the repeated occurrence in time of an event, according to given deterministic or stochastic laws. Often linearity holds, and one is faced with signal models of the form

$$s[n] = w[n] \star r[n] \tag{3-27}$$

where $s[n]$ denotes the received signal, $w[n]$ a sequence associated with the physical event, and $r[n]$ an impulse train, of the form

$$r[n] = \sum_{k=1}^{M} r_k \delta[n - n_k], \qquad n_{k+1} > n_k, \quad r_k \neq 0 \tag{3-28}$$

where (r_k, n_k), $k = 1, \ldots, M$, denote the arrival times and amplitudes associated with the occurrences of the event. In general, the pulse $w[n]$ is much shorter then $r[n]$.

For example, in speech analysis [4] $w[n]$ represents the combined glottal pulse/vocal tract effect and $r[n]$ represents either a periodic impulse train (voiced speech) or a random series (unvoiced speech). In teleseismic applications [8], $w[n]$ represents the seismic wavelet, while $r[n]$ is a random impulse train associated with the structure of the earth's crust. We shall then derive a number of properties regarding the complex cepstrum structure of pulses $w[n]$ and impulse trains $r[n]$.

PROPERTY 6

The complex cepstrum $\hat{w}[n]$ of a pulse $w[n]$ whose spectrum is smooth tends to be concentrated around low-time values. This property derives from the fact that a sequence $w[n]$ with a smooth spectrum $|W(e^{j\omega})|^2$ is bound to have relatively broad-band poles and zeros. In terms of Property 1, this means that $|\alpha^n/n|$ decays rather rapidly with n and Property 6 thus follows.

Let us define in general, a *periodic impulse train* $r[n]$ with period $T > 1$, as one for which the interarrival times are multiples of T, that is,

$$n_k = n_{k-1} + l_k T, \qquad k = 2, \ldots, M, \quad \forall\, l_k > 0$$

It can then be shown that:

PROPERTY 7

The complex cepstrum $\hat{r}[n]$ of a periodic impulse train $r[n]$ is also a periodic impulse train, with the same period, that is,

$$\hat{r}[n] = \sum_{k=-\infty}^{\infty} \hat{r}[k]\delta[n - kT] \tag{3-29}$$

Note that the result above holds, in general, no matter what the impulse amplitudes r_k are. Thus, the complex cepstrum of an *aperiodic* impulse train, that is, an impulse train for which the interarrival times are noncomensurable ($T = 1$), will in general be nonzero for all times.

In many physical situations the amplitudes of the impulse train $r[n]$ obey certain deterministic or probabilistic laws. Often it is possible to derive a priori the fact that an impulse train is minimum-phase. Using the recursive relationship for minimum-phase sequences (Property 4), the following can be shown.

PROPERTY 8

Consider a minimum-phase impulse train $r[n]$ for which the first interarrival time is N, so that

$$n_2 - n_1 = N, \qquad n_k \text{ arbitrary}, \quad k > 2$$

Then the complex cepstrum is zero, for $0 < n < N$. In general, $\hat{r}[n]$ is nonzero, only at times $0, n_2 - n_1, n_3 - n_1, \ldots, n_M - n_1$, as well as at all positive linear combinations of these times. A similar property can be derived for maximum-phase impulse trains:

PROPERTY 9

Consider a maximum-phase impulse train $r[n]$ for which the last interarrival time is N, so that

$$n_M - n_{M-1} = N, \qquad n_k \text{ arbitrary}, \quad k < M - 1$$

Then the complex cepstrum is zero for $-N < n < 0$. In general, $r[n]$ is nonzero only at times $0, -(n_M - n_{M-1}), -(n_M - n_{M-2}), \ldots, -(n_M - n_1)$ as well as at all positive linear combinations of these times.

Properties 7, 8, and 9 are extremely useful in a number of applications since they enable us to predict the cepstral characteristics of a large class of impulse trains given a minimum amount of knowledge of the corresponding time structure.

The only class that remains to be discussed is the class of aperiodic mixed-phase impulse trains. In this case the implicit relationship of equation (3-23b) is of no help in attempting to

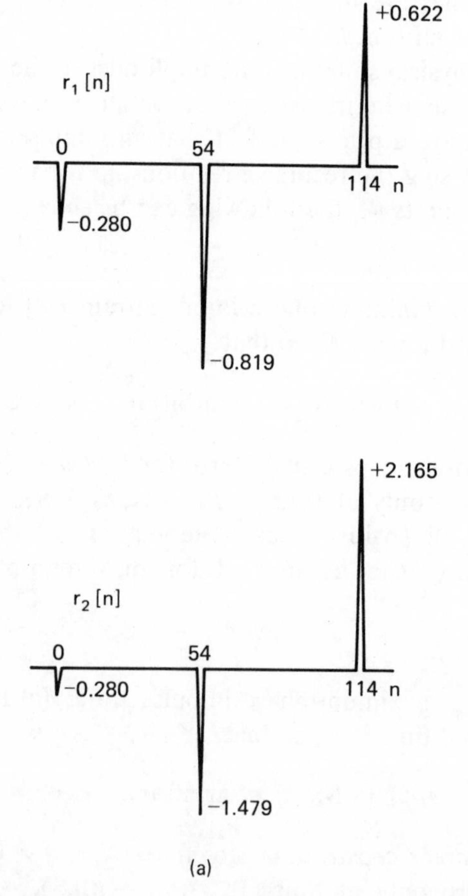

Figure 3.5 Sensitivity of the cepstral structure of impulse trains to time-domain amplitude perturbations: (a) two impulse trains, $r_1[n]$ and $r_2[n]$

establish a link between the time and the cepstral structures, for such class of sequences.

We attempted to establish such a link by running a large number of experiments, where a set of random impulse trains was generated, all with the same arrival times but with slight amplitude variations, and the corresponding cepstra were computed and compared. We illustrate this approach in Figure 3.5. Figure 3.5a

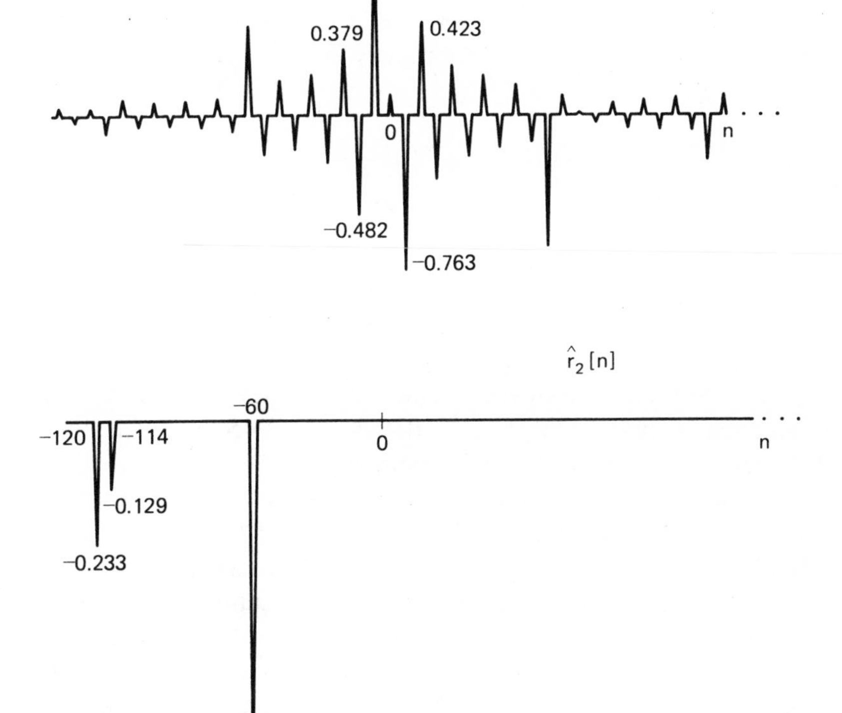

Figure 3.5 (Cont.) (b) complex cepstra $\hat{r}_1[n]$ and $\hat{r}_2[n]$

depicts two impulse trains $r_1[n]$ and $r_2[n]$, which share the same set of arrival times. Their complex cepstra $\hat{r}_1[n]$ and $\hat{r}_2[n]$ are shown in Figure 3.5b.

Two experimental findings were observed repeatedly. The first one is that the cepstral structure of an impulse train may be very sensitive to minor changes in the impulse amplitudes. The second is that the cepstra of mixed-phase impulse trains often exhibit a rather elaborate structure with both low-time and high-time components, which, within the limits imposed by our present understanding of the cepstral mapping, offer no clues regarding the structure of the corresponding time structure, and vice versa.

These two characteristics should be kept in mind throughout this book, as they play a major role in shaping the role of homomorphic systems in seismic data analysis.

III.7 HOMOMORPHIC SIGNAL ANALYSIS

The analysis of a convolution of two sequences into its components by means of a homomorphic system for convolution depends exclusively on our ability in designing a linear operator in the cepstral domain which is able to recover the cepstrum of either component from their sum. Such a linear system must be designed in practice with little detailed knowledge of the time-domain structure of the signal components.

As a very general comment, we can say that the type of signals for which homomorphic analysis has thus far been found useful are those for which the complex cepstra of the signal components, in some sense, do not overlap. An obvious example of this is the analysis of a signal into its minimum-phase and maximum-phase components. In fact, homomorphic filtering is, at the present, the only general technique available for such a purpose, with the exception of z-transform factorization methods, which are reliable only for sequences with lengths not exceeding, say, 256 samples [11].

Another example where there is, approximately, no overlapping between the cepstra of the signal components, arises in the analysis of voiced speech, where one of the components, the vocal-tract impulse response, has a smooth spectrum and has therefore the corresponding cepstrum concentrated in the low-time region, whereas the other component, a pitch periodic impulse train, has a complex cepstrum that exhibits the same periodicity and has, therefore, no contributions in the low-time region. Thus, the type of linear systems that has been found useful in separating the complex cepstra of the convolved components are frequency-invariant systems, of the form

$$\hat{y}[n] = L[x[n]] = l[n]\hat{x}[n] \tag{3-30}$$

which essentially attempt to set to zero the intervals in the cepstral domain which are thought to be associated with the unwanted signal components. These systems are commonly referred to as *cepstral windows* or *cepstral gates*. Figure 3.6 depicts such a class of systems.

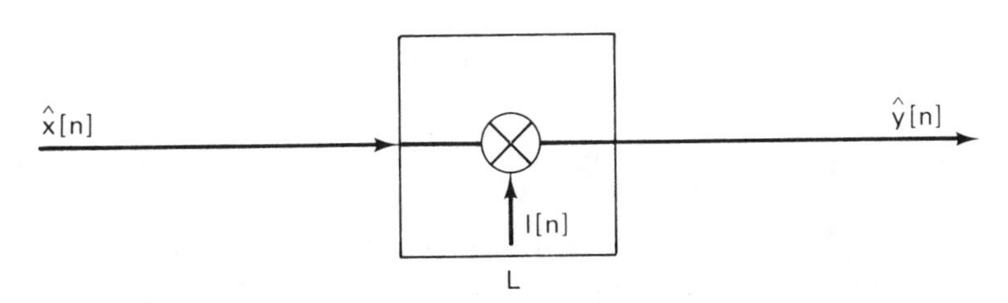

Figure 3.6 Frequency-invariant operators in the cepstral domain

In general, the use of different types of linear systems in the cepstral domain is, of course, theoretically possible. For example, one might want to apply different frequency-invariant filters on the magnitude and phase of a signal. Thus, letting $l_1[n]$ and $l_2[n]$ be even windows, we may independently window the even and odd components as

$$\hat{y}_e[n] = l_1[n]\hat{x}_e[n] \tag{3-31a}$$

$$\hat{y}_o[n] = l_2[n]\hat{x}_o[n] \tag{3-31b}$$

and then add both results as

$$\hat{y}[n] = \hat{y}_e[n] + \hat{y}_o[n] \tag{3-32}$$

Or, one may not want to use a frequency-invariant filter at all. For example, one might wish to design a homomorphic system which maps every input $x[n]$ into a minimum-phase output $y[n]$ such that both signals have exactly the same phase, that is, $\arg|X(e^{j\omega})| = \arg|Y(e^{j\omega})|$. In that case, the required linear system is described by

$$\hat{y}[n] = [\hat{x}[n] - \hat{x}[-n]]u[n-1] \tag{3-33}$$

However, within the framework of deconvolution by homomorphic filtering, the class of linear filters illustrated in Figure 3.6 has been, thus far, the only one found to be useful. A number of basic types of windows have been incorporated in homomorphic filters, depending on the cepstral structure of the signal components. In speech analysis, for example, the use of symmetric low-time windows with a total width less than twice the pitch period have been found most appropriate for the recovery of the vocal-tract pulse. Other types of commonly used windows are high-time, notch, and comb windows.

Homomorphic deconvolution—by this we mean, precisely, the filtering of a composite signal through a homomorphic system, with the goal of recovering one of its components at the output—involves two basic assumptions. The first one is that the signal components occupy disjoint intervals in the cepstral domain. This assumption will hold for a large class of signal models, where one of the components is a pulse with smooth spectrum, the other being a periodic, or minimum-phase, or maximum-phase impulse train. This assumption will not hold in general when one of the components is a mixed-phase impulse train.

The second assumption involves the design of the cepstral window. This usually simply requires the estimation of a few parameters, such as the period of a periodic impulse train or the first interarrival time of a minimum-phase train. This estimation can often be done either by physical considerations or by direct estimation in the cepstral domain.

Within this framework, homomorphic deconvolution compares very favorably with linear inverse filtering methods, in that it does not require knowledge of one of the signal components.

III.8 FULL-BAND HOMOMORPHIC SYSTEMS

Thus far we have restricted the input signals to a full-band homomorphic system to be stable sequences with stable inverses. Although the first requirement is usually guaranteed by the inherent stability of the physical processes, the second requirement is often not satisfied, either because the signals we are dealing with have bandpass characteristics or because they might have isolated zeros on the unit circle. For example, in many applications, the bias of the data is automatically removed, thus introducing a zero at dc.

In attempting to develop a class of homomorphic systems for convolution which can handle signals without stable inverses, we shall first consider a simple generalization of the complex logarithmic characteristic system of equation (3-1). If we define

$$\hat{X}(z) = \log\,(X(\alpha^{-1}Z)), \qquad \alpha > 0 \tag{3-34}$$

we may derive, using arguments similar to those of Section III.3, that the input space for such a characteristic system is restricted to sequences $x[n]$ such that the z-transforms $X(z)$ and $X^{-1}(z)$ are analytic on the circular contour of radius α^{-1}. Thus, the inputs are no longer required to have stable inverses, as long as their z-transforms converge on some annular region in the z-plane.

We conclude, therefore, that low-pass, high-pass, bandpass signals, and in general any signals whose Fourier transforms are zero over one or more frequency *bands* must still be excluded from the input space to the characteristic system defined by equation (3-33), since the z-transform of such signals are only defined on the unit circle. We shall emphasize this fact by referring to the class of homomorphic systems with such characteristic systems as the class of *full-band homomorphic systems* and to the class of signals these systems can handle as *full-band signals.*

Noting that

$$X(\alpha^{-1}z) = Z(\alpha^{n}x[n]) \tag{3-35}$$

the requirements at the input of a full-band homomorphic system may be formulated in terms of the exponentially weighted input sequence $\alpha^n x[n]$, which must now be a SSI sequence. One can represent such characteristic system as in Figure 3.7, where the

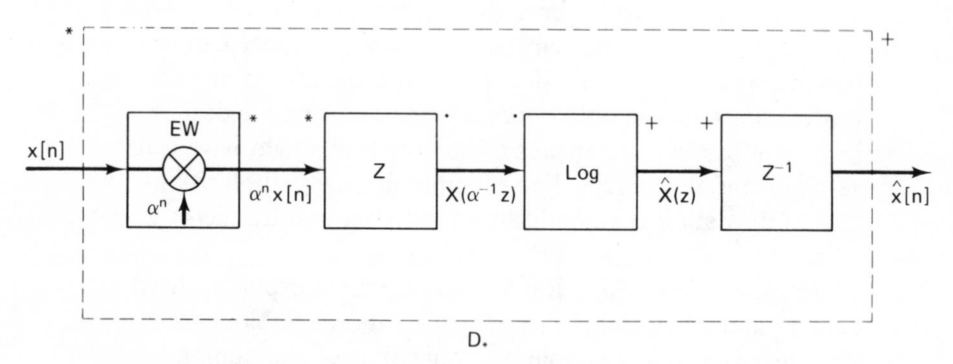

Figure 3.7 Generalized complex logarithmic characteristic system

system EW denotes the exponential weighting operation on the input $x[n]$. Of course, the corresponding inverse characteristic system $D_\star^{-1}$ will include the system $(\mathrm{EW})^{-1}$, which corresponds to an unscaling operation by means of an exponential weighting by α^{-n}.

III.9 EXPONENTIAL WEIGHTING

The choice of the weighting α introduces an extra degree of freedom in homomorphic signal analysis. In fact, exponentially weighting has the effect of radially scaling the poles and zeros of the z-transforms of the inputs. Thus, for example, a mixed-phase impulse train may be converted into a minimum-phase impulse train by exponential weighting with a small-enough value of α so that all zeros lying outside the unit circle are moved inside. This is often of interest in situations where one of the signal components is a mixed-phase impulse train. The weighted train, being minimum-phase, has a cepstral structure that can now be easily determined.

This technique, first suggested by Schafer [12], has been widely used in seismic deconvolution. A critical assessment of this practice will be presented in Chapter VIII.

III.10 SUMMARY

A class of homomorphic systems for convolution can be defined using a complex logarithmic characteristic system whose output is referred to as the complex cepstrum of the input. Such a class of systems, denoted as full-band homomorphic systems, is appropriate for the processing of full-band sequences.

The structure of the complex cepstrum is such that for a number of applications, the complex cepstra of the signal components tend to occupy disjoint time intervals. This tendency enables the separation of the signal components by cepstral gating. Within such a context, homomorphic deconvolution compares very favorably with linear inverse filtering, in that the former does not require the detailed knowledge of one of the signal components.

IV

Homomorphic Systems for Convolution: Bandpass Systems

IV.1 INTRODUCTION

In many applications, the signals one wishes to analyze have intrinsic bandpass characteristics. For example, owing to the low-pass filtering of the earth, seismic data are often devoid of significant high-frequency components. Furthermore, at very low frequencies the signal-to-noise ratio is very poor. As a consequence, we may wish to represent the data components which are pertinent to the phenomena under observation in terms of idealized bandpass signal models and to process the data accordingly. In fact, seismic data are, in general, bandpass-filtered, prior to sampling and recording, to discriminate against the noise.

The homomorphic analysis of such idealized bandpass signal models *cannot* be accomplished by means of full-band homomorphic systems. In fact, the analysis cannot be performed on the unit circle, since then the logarithm would become unbounded in the frequency bands with zero energy. Neither can it be performed off the unit circle, since the z-transform of such signals does not converge anywhere in the z-plane, but on the unit circle. Thus, the exponential weighting procedure used to remove the zeros of a signal off the unit circle cannot be applied to the class of bandpass signals.

We are thus motivated to search for a class of homomorphic systems for convolution which is appropriate for the analysis of bandpass signals. A solution involves a restriction on the domain of the logarithmic mapping, to encompass only the passband of the input. This approach may be conveniently formulated in terms of a frequency scaling operation that shifts and stretches the signal's passband to occupy the entire frequency band. The result of this operation is then a full-band sequence, which may be analyzed using full-band homomorphic systems.

The discussion above was based upon idealized bandpass signal representations. Of course, real data are always bound to have some residual out-of-band noise. Furthermore, real data will always have z-transforms that converge on some annular region of the z-plane. In particular, real data can only have a countable number of zeros on the unit circle. Thus, it becomes possible to use exponential weighting to remove such zeros off the unit circle and therefore process any type of real data through a full-band homomorphic system. In fact, seismic data are currently analyzed through such a class of systems.

It is important, however, to realize that such processing relies on the presence of the out-of-band noise, which we thought to eliminate in the first place, by bandpass filtering. This approach is thus inherently ill-conditioned, and it often leads to nonrobust bandpass signal analysis.

IV.2 THE CHARACTERISTIC SYSTEM

As discussed in Chapter II, our approach to homomorphic system synthesis is characterized by first seeking candidates for a characteristic system and then imposing on them the necessary and sufficient continuity and invertibility requirements. The key to the selection of a characteristic system for the homomorphic analysis of bandpass-filtered sequences is to realize that any bandpass-filtered signal can be mapped into a full-band signal, by appropriate frequency scaling, as illustrated in Figure 4.1.

We shall henceforth refer to such an operation as *bandpass mapping*. This suggests the synthesis of the characteristic system as the cascade of such mapping with the complex logarithmic

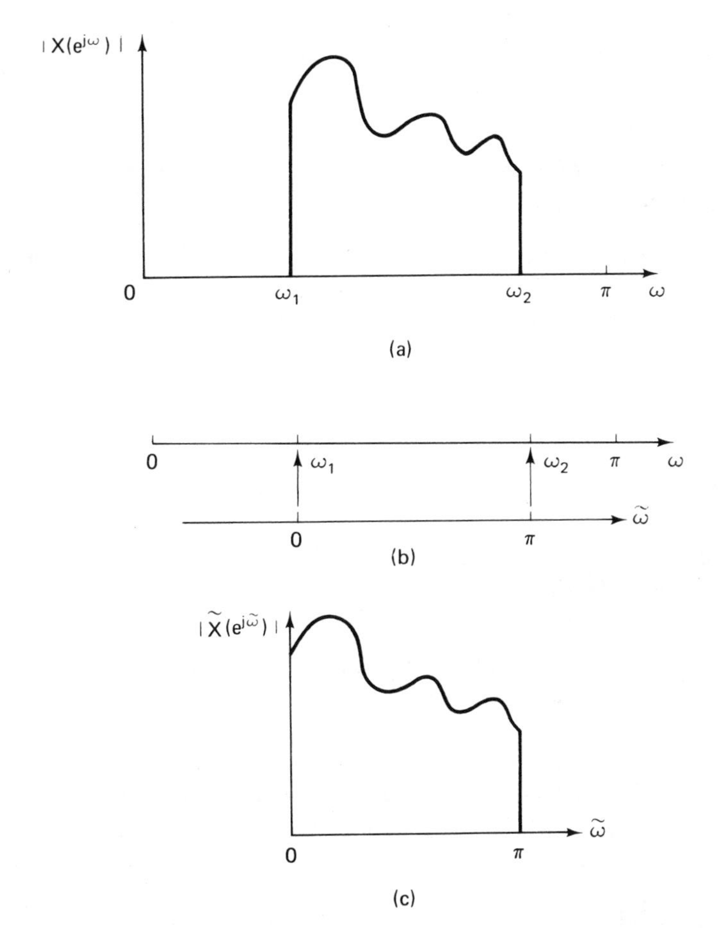

Figure 4.1 Bandpass mapping: (a) magnitude spectrum of bandpass signal $x[n]$; (b) frequency transformation; (c) magnitude spectrum of full-band signal $\tilde{x}[n]$

system used in the homomorphic analysis of full-band signals. In this way, the burden of proving if such a cascade has the required properties of a characteristic system is now reduced to proving that an invertible bandpass mapping can be defined, which is, itself, homomorphic for convolution.

This result shall be demonstrated next. Let S_B denote a vector space with convolution as vector addition, formed by all stable sequences $x[n]$, satisfying

$$X(e^{j\omega}) = \begin{cases} X(e^{j\omega}) \neq 0, & \omega_1 < |\omega| < \omega_2 \\ X_R(e^{j\omega}) \neq 0, & |\omega| = \omega_1, \omega_2 \\ 0, & \text{elsewhere} \end{cases} \tag{4-1}$$

where ω_1 and ω_2 denote the cutoff frequencies, which are common to all sequences in S_B.

Let BP denote the bandpass mapping operation defined as

$$\tilde{x}[n] = \mathrm{BP}\{x[n]\}, \qquad \forall\ x[n] \in S_B \tag{4-2}$$

where $x[n]$ is a real, stable sequence such that

$$\tilde{X}(e^{j\tilde{\omega}}) = X(e^{j\omega}), \qquad 0 \leq |\tilde{\omega}| \leq \pi \tag{4-3}$$

where

$$\tilde{\omega}(\omega) = \pi \frac{\omega - \omega_1}{\omega_2 - \omega_1} \qquad \omega_1 \leq |\omega| \leq \omega_2 \tag{4-4}$$

Let S denote the set of all sequences $\tilde{x}[n]$. Then we can show that the following theorem is true.

THEOREM

The bandpass-mapping BP is an invertible homomorphic mapping between S_B and S, with convolution as input and output operations.

PROOF

Let $\tilde{x}[n] = \mathrm{BP}\{x[n]\}$. Consider the change of variables

$$\omega' = \omega_1 + \tilde{\omega}\frac{\omega_2 - \omega_1}{\pi}, \qquad \omega_1 \leq \omega' \leq \omega_2 \tag{4-5}$$

Define a signal $x'[n]$ such that

$$X'(e^{j\omega'}) = \begin{cases} \tilde{X}(e^{j\tilde{\omega}}), & \omega_1 \leq \omega' \leq \omega_2 \\ 0, & \text{elsewhere} \end{cases} \tag{4-6}$$

Then $x'[n] \in S_B$. Furthermore, from equation (4-3), we conclude that

$$X'(e^{j\omega'}) = X(e^{j\omega}), \qquad \forall\ \omega = \omega' \tag{4-7}$$

Thus, $x'[n] = x[n]$ and the transformation defined by equation (4-5) corresponds to the inverse of BP, BP^{-1}.

Let $x_1[n], x_2[n] \in S_B$. Then

$$\begin{aligned} \mathrm{BP}\{x_1[n] \star x_2[n]\} &= \mathrm{IFT}[\tilde{X}_1(e^{j\tilde{\omega}})\tilde{X}_2(e^{j\tilde{\omega}})] \\ &= \mathrm{IFT}[\tilde{X}_1(e^{j\tilde{\omega}})] \star \mathrm{IFT}[\tilde{X}_2(e^{j\tilde{\omega}})] \\ &= \mathrm{BP}\{x_1[n]\} \star \mathrm{BP}\{x_2[n]\} \end{aligned} \tag{4-8}$$

The mapping BP is thus homomorphic for convolution.

We shall now formally define a canonical *bandpass homomorphic system* for convolution as a canonical homomorphic system whose characteristic system $D_\star$ is defined as

$$\hat{x}[n] = D_\star[x[n]] \tag{4-9a}$$

where

$$\hat{X}(z) = \log \tilde{X}(z) \tag{4-9b}$$

and

$$\tilde{x}[n] = \mathrm{BP}[x[n]] \tag{4-9c}$$

The characteristic system $D_\star$ for bandpass homomorphic systems is illustrated in Figure 4.2.

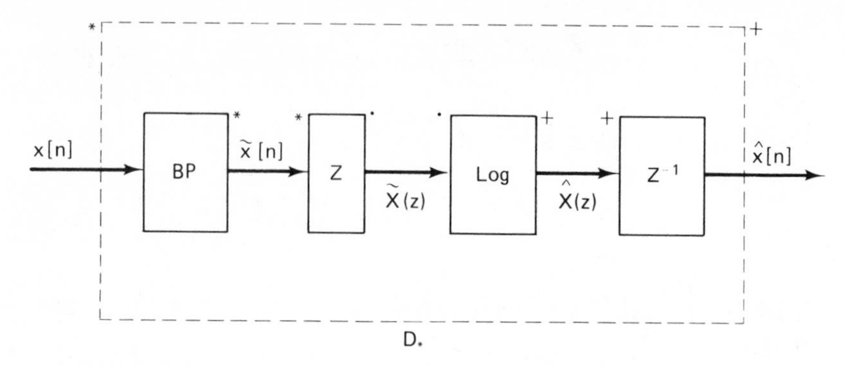

Figure 4.2 Characteristic system $D_\star$ for bandpass homomorphic systems

IV.3 INPUT NORMALIZATION

Let us now take the final step in defining the class of bandpass homomorphic systems, by finding the normalization mapping N_B, associated with the choice of characteristic system $D_\star$ made above.

The need for input normalization can be intuitively understood in terms of removing any phase discontinuities between positive and negative frequency passbands. If these discontinuities were kept, they would generate transient responses to the frequency-shift-invariant filters used in the cepstral domain. These transients, originating from localized discontinuities, would spread out through the signal's passband, seriously undermining our analysis efforts.

Input normalization corresponds, then, to an organization of the inputs into classes of signals that have equivalent frequency structures, except with regard to the size of the phase discontinuities between the passbands. The role of the normalization N_B is thus to map to the input of $D_\star$ that one sequence in each class which has no phase discontinuities at all.

The mathematical restrictions on the inputs to $D_\star$ can be essentially derived from the continuity requirements at the input to the complex logarithmic mapping. That is, letting $s_{\tilde{x}}$ and $r_{\tilde{x}}$

denote the polarity and lag of $\hat{x}[n]$, one must have, at the output of BP that $s_{\hat{x}} = +1$ and $r_{\hat{x}} = 0$. It follows that at the input to $D_{\star}$, we must require that

$$s_{\hat{x}} = \text{SIGN}\,[\tilde{X}(e^{j0})] = \text{SIGN}\,[X_R(e^{j\omega_1})] = +1 \tag{4-10a}$$

and

$$\begin{aligned} r_{\hat{x}} &= \frac{1}{\pi}\int_0^{\pi}\left[\frac{d\arg|\tilde{X}(e^{j\tilde{\omega}})|}{d\tilde{\omega}}\right]d\tilde{\omega} \\ &= \frac{1}{\omega_2-\omega_1}\int_{\omega_1}^{\omega_2}\left[\frac{X_R(e^{j\omega})X_I'(e^{j\omega}) - X_I(e^{j\omega})X_R'(e^{j\omega})}{|X(e^{j\omega})|^2}\right]d\omega = 0 \end{aligned} \tag{4-10b}$$

The required normalization mapping within S_B then turns out to be

$$N_B[X(e^{j\omega})] = \begin{cases} (s_{\hat{x}}e^{-j\omega_1 r_{\hat{x}}})X(e^{j\omega})e^{j\omega r_{\hat{x}}}, & \omega > 0 \\ (s_{\hat{x}}e^{+j\omega_1 r_{\hat{x}}})X(e^{j\omega})e^{j\omega r_{\hat{x}}}, & \omega < 0 \end{cases} \tag{4-11}$$

It can easily be seen that the normalization N_B reduces to the mapping N of equation (3-15) when $\omega_1 = 0$ and $\omega_2 = \pi$. Using arguments similar to those used in Chapter III, it can be shown that the mapping N_B satisfies the required conditions to be cascaded with a canonical homomorphic system for convolution. In particular, N_B is homomorphic for convolution.

Let us conclude our remarks on input normalization by noting that the vector space S_B has within it an implicit normalization, as seen in equation (4-1), where it is required that the phase of $X(e^{j\omega})$ at both band edges be a multiple of π.

This condition derives from the need of avoiding loss of information through band-edge aliasing resulting from the definition of $\tilde{\omega}(\omega)$ in equation (4-4). There is, however, no loss of generality in such an approach, since one could as easily define the frequency scaling in such a way that no aliasing would ever occur:

$$\tilde{\omega}(\omega) = \pi\frac{\omega-\omega_1}{\omega_2-\omega_1}, \qquad \omega_1 < |\omega| < \omega_2$$

But then, for the BP mapping to be well defined, we would have to impose that

$$\tilde{X}(e^{j\tilde{\omega}}) = 0, \qquad \tilde{\omega} = 0, \pi$$

Such BP systems would also be invertible and homomorphic for convolution. However, $\tilde{x}[n]$ would have two zeros on the unit circle, and the use of exponential weighting would be required prior to the complex logarithmic transformation. Since, in practice, the implementation of either BP mappings will certainly give rise to some aliasing, the subtle mathematical distinction between both approaches becomes irrelevant.

IV.4 BANDPASS MAPPING

Seismograms may be represented as a sum of wavelets, with different strengths and arrival times. These models, which can be theoretically derived in continuous time, are commonly assumed to hold in sampled form, for a wide range of sampling rates. In this section we shall use the formalism of bandpass mapping to show that these same models may also be accommodated, to represent bandpass-mapped seismic data. The mapping BP satisfies two important properties:

1. BP is a homomorphic system for convolution, mapping S_B onto S.
2. BP is a linear time-varying system, mapping S_B onto S.

Let $b[n] \in S_B$ denote an ideal bandpass filter, defined as

$$B(e^{j\omega}) = \begin{cases} 1, & \omega_1 \leq |\omega| \leq \omega_2 \\ 0, & \text{elsewhere} \end{cases} \tag{4-12}$$

Then, since for any $x[n] \in S_B$,

$$x[n] = x[n] \star b[n] \tag{4-13a}$$

it follows that

$$x[n] = \sum_{k=-\infty}^{\infty} x[k]b[n-k] \tag{4-13b}$$

Letting $\tilde{b}[n; k]$ denote the response of BP to the input $b[n-k]$, we may write

$$\tilde{x}[n] = \sum_{k=-\infty}^{\infty} x[k]\tilde{b}[n; k] \tag{4-14}$$

The set of sequences $\tilde{b}[n; k]$ satisfy a number of interesting properties:

1. $\tilde{b}[n; r] = \tilde{b}[n; k] \star \tilde{b}[n; r-k], \ \forall\ r, k.$
2. $|\tilde{B}(e^{j\omega}; k)| = 1, \ \forall\ k, \omega.$
3. $\tilde{b}[n; lM + r] = \tilde{b}[n - lM; r], \ \forall\ l, r,$
   where $M = \text{l.c.m.}\ (2\pi/\omega_1, 2\pi/\omega_2).$

In general, it can be shown that

$$\tilde{b}[n; k] = \cos\left(\frac{n\pi}{2} - k\frac{\omega_2 + \omega_1}{2}\right) \frac{\sin\left(\frac{n\pi}{2} - k\frac{\omega_2 - \omega_1}{2}\right)}{\frac{n\pi}{2} - k\frac{\omega_2 - \omega_1}{2}} \tag{4-15}$$

Thus, all sequences $\tilde{b}[n; k]$ have the same envelope, which is simply a cosine amplitude-modulated $\sin x/x$ function, sampled with a $\pi/2$ interval. This result is independent of the values of the cutoff frequencies ω_1 and ω_2. Because of the fast decaying characteristics of the $\sin x/x$ function, the most significant samples of each $\tilde{b}[n; k]$ will correspond to those resulting from the main lobe of $\sin x/x$. Since the width of this lobe is π, each $\tilde{b}[n; k]$ will essentially consist of a sharp burst with a time width of 2 samples, centered around time $n_0 \simeq +[(\omega_2 - \omega_1)/\pi]k$.

Consider now a bandpass-filtered signal $s[n] \in S_B$ of the form

$$s[n] = w[n] \star r[n] \tag{4-16}$$

where $w[n] \in S_B$ represents a pulse and $r[n]$ an impulse train of the form

$$r[n] = \sum_{k=1}^{M} r_k \delta[n - n_k] \tag{4-17}$$

where the r_k's and n_k's might represent, for example, the amplitudes and arrival times of the various wavelets in a seismic signal.

It therefore follows that

$$\tilde{s}[n] = \tilde{w}[n] \star \tilde{r}[n] \tag{4-18}$$

where we have defined $\tilde{r}[n]$ as

$$\tilde{r}[n] = \text{BP}\{r[n] \star b[n]\} \tag{4-19}$$

Then, from equation (4-14), $r[n]$ satisfies

$$\tilde{r}[n] = \sum_{k=1}^{M} r_k \tilde{b}[n; n_k] \tag{4-20}$$

and, if the wavelet arrival times Δ_k, defined as

$$\Delta_k \triangleq n_k - n_{k-1}, \qquad k = 2, \ldots, M$$

satisfy

$$\Delta_k > \frac{2\pi}{\omega_2 - \omega_1}, \ \forall\ k \tag{4-21}$$

then each arrival of $\tilde{w}[n]$ in equation (4-18) is clearly resolved, in terms of the impulsive burst $\tilde{b}[n; n_k]$.

Consider now a BP system for which $\omega_1 = 0$. Then BP simply corresponds to a sampling-rate reduction of the original sequence $s[n]$. Different sampling rates amount to different choices of ω_2, therefore to different degrees of arrival-time resolution. As is clear from equation (4-21), this result is valid in general when $\omega_1 \neq 0$, the arrival-time resolution being determined by the width of the passband $\omega_2 - \omega_1$. There is, of course, a certain amount of pulse distortion involved in these transformations. This is due to the time-varying nature of the mapping and is translated into the dependence of the shape of the impulsive burst $\tilde{b}[n; n_k]$ on n_k. However, this distortion may usually be assumed to be negligible when compared with the deviations of real data from our idealized models.

IV.5 BANDPASS SIGNAL ANALYSIS

The homomorphic analysis of bandpass-filtered signals can be conceptually divided into three steps:

1. Bandpass mapping through BP.
2. Full-band homomorphic analysis.
3. Inverse bandpass mapping through $(\text{BP})^{-1}$,
   as depicted in Figure 4.3.

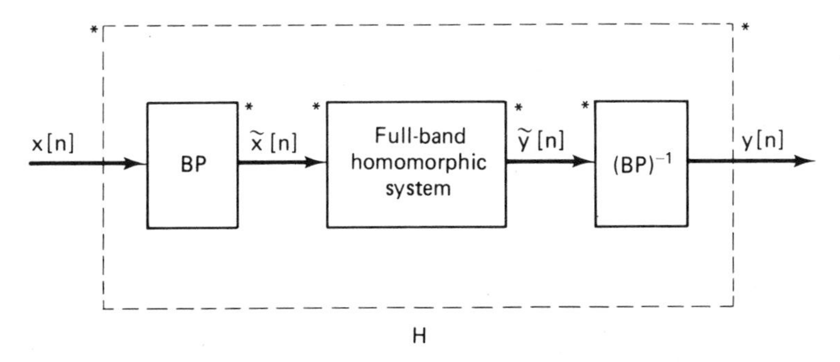

Figure 4.3 Bandpass homomorphic system

Thus, the design of the linear component in the cepstral domain is subject to considerations similar to those in Section III.7, but now relative to the bandpass-mapped inputs $\tilde{x}[n]$, not to the inputs $x[n]$.

IV.6 SUMMARY

The class of homomorphic bandpass systems is matched to the analysis of signals whose Fourier transform can be represented as the product of two components, but only within a given passband.

Bandpass homomorphic systems may be defined as the cascade of a bandpass mapping system with a full-band homomorphic system followed by the inverse bandpass mapping system.

The first system corresponds to a frequency scaling operation that shifts and stretches the passband of the input signals to occupy the entire frequency band. The resulting output is then a full-band sequence, which can be analyzed using a full-band homomorphic system. The inverse bandpass mapping rescales the frequency axis so that the resulting outputs regain the original bandpass characteristics.

V

Homomorphic Systems for Convolution: Implementation

V.1 INTRODUCTION

We have shown in Chapters III and IV that the class of homomorphic bandpass systems can be represented as the cascade of subsystems, illustrated in Figure 5.1.

This chapter is concerned with the implementation of homomorphic systems on a digital computer. Since digital computers perform finite computations, we are limited to finite-length input sequences, and we can compute the Fourier transforms at only a finite number of points. That is, instead of using the Fourier transform (FT), we must use the discrete Fourier transform (DFT).

The principal computational step in a DFT implementation of homomorphic system is the evaluation of the samples of the continuous complex logarithm, LOG. The problem lies in the

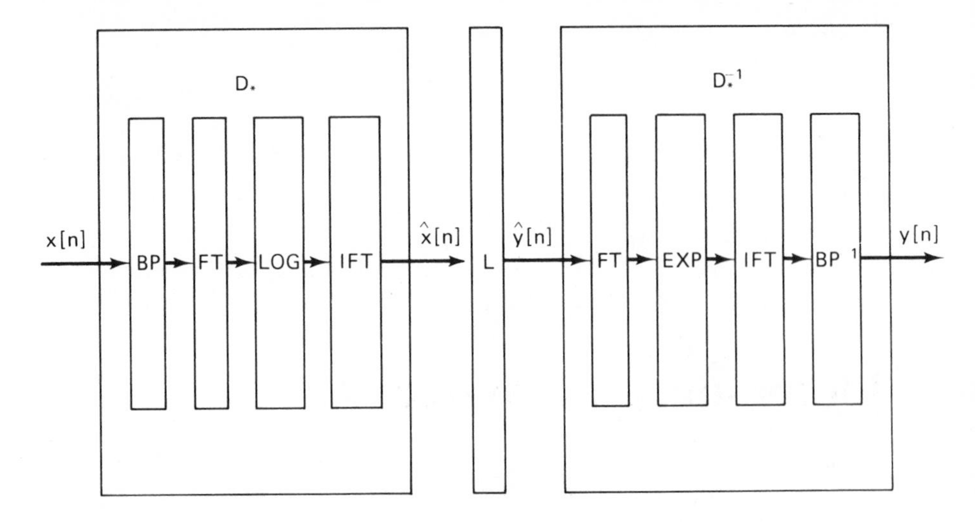

Figure 5.1 Representation of homomorphic bandpass systems

evaluation of the continuous phase. This evaluation, which is commonly referred to as *phase unwrapping*, has been thoroughly investigated in this book. As a result, an unwrapping technique based on an adaptive numerical integration scheme is proposed [13] which is not only highly reliable but which is matched to DFT realizations by fast Fourier transform (FFT) algorithms, thus allowing fast and efficient homomorphic filtering on general- and on special-purpose hardware.

Finally, the implementation of the bandpass-mapping system BP and $(BP)^{-1}$ can be efficiently and reliably done using digital interpolation and decimation in conjunction with strictly linear-phase FIR low-pass filters [14].

V.2 DFT IMPLEMENTATION

Let $x[n]$ denote a full-band sequence. Then the complex cepstrum $\hat{x}[n]$ is given by

$$\hat{x}[n] = \text{IFT}\{\hat{X}(e^{j\omega})\} = \frac{1}{2\pi}\int_{-\pi}^{\pi} \log\,[s_x e^{+j\omega r_x} X(e^{j\omega})]e^{j\omega n}\,d\omega \qquad (5\text{-}1)$$

Since, in practice, we are limited to finite-length input sequences, we shall take that into consideration by assuming $x[n]$ to exist in the interval $0 \leq n \leq N-1$.

Motivated by the tremendous advantages of FFT algorithms, we shall seek implementations using the DFT. Thus, instead of equation (5-1), we have the computational realization

$$\hat{x}_p[n] = \frac{1}{N} \sum_{k=0}^{N-1} \hat{X}(e^{j2\pi k/N}) e^{j2\pi kn/N} \tag{5-2a}$$

where $\hat{X}(e^{j2\pi k/N})$ is the sampled complex logarithm

$$\hat{X}(e^{j2\pi k/N}) = \log \left\{ s_x e^{(+j2\pi k/N) r_x} \sum_{l=0}^{N-1} x[l] e^{-j2\pi kl/N} \right\} \tag{5-2b}$$

Since the complex cepstrum is computed from samples of its Fourier transform, time aliasing of the complex cepstrum results. That is,

$$\hat{x}_p[n] = \sum_{k=-\infty}^{\infty} \hat{x}[n + kN], \qquad \forall\, n \tag{5-3}$$

The complex cepstrum $\hat{x}[n]$ has, in general, infinite duration. Thus, a certain amount of aliasing will always occur. Let us rewrite equation (5-3) as

$$\hat{x}_p[n] = \hat{x}[n] + \hat{x}_a[n], \qquad |n| < N/2 \tag{5-4}$$

where $\hat{x}_a[n]$ denotes the aliasing component in the interval $|n| < N/2$, that is,

$$\hat{x}_a[n] = \sum_{\substack{k=-\infty \\ k \neq 0}}^{\infty} \hat{x}[n + kN], \qquad |n| < N/2 \tag{5-5}$$

We have shown that $\hat{x}[n]$ decays faster than an exponential sequence. Thus, it is expected that the aliasing component $\hat{x}_a[n]$ can be reduced by appending zeros to the data $x[n]$, before computing the DFT, in order to increase the total length N. This corresponds to a finer spectral sampling on $\hat{X}(e^{j\omega})$.

The choice of a given transform size is associated then with a judgment on how much cepstral aliasing one is willing to tolerate. As with all problems involving aliasing phenomena, such judgment is very much application-dependent.

The DFT implementation of the inverse cepstral mapping yields

$$y_p[n] = \frac{1}{N} \sum_{k=0}^{N-1} Y(e^{j2\pi k/N}) e^{j2\pi kn/N} \tag{5-6a}$$

where $Y(e^{j2\pi k/N})$ is the sampled complex exponential

$$Y(e^{j2\pi k/N}) = \exp \left\{ \sum_{n=-[(N-1)/2]}^{(N-1)/2} \hat{y}[n] e^{-j2\pi kn/N} \right\} \tag{5-6b}$$

where, again, $y_p[n]$ is an aliased version of $y[n]$ with period N, which might accurately represent $y[n]$ to the extent that $y[n]$ is a finite sequence of length not greater than N.

V.3 THE UNWRAPPED PHASE

The DFT implementation discussed above requires the evaluation of samples of the continuous complex logarithm as in equation (5-2b). In particular, we need to determine samples of the continuous-phase function of the normalized input.

As we discussed in Chapter III, the input normalization is associated with the proper definition of the continuous phase. The normalization of the signal polarity can be easily done. However, the signal lag is, in general, not known a priori. It is thus convenient in practice to implement the required input normalization while in the process of evaluating the continuous phase. Thus, using arguments similar to those of Chapter III, we define a phase function $\arg_u |X(e^{j\omega})|$ associated with the nonnormalized input $x[n]$ as an integral of the form

$$\arg_u |X(e^{j\omega})| = \int_0^{\omega} \left[\frac{d}{d\eta} \arg'_u |X(e^{j\eta})| \right] d\eta, \qquad 0 \leq \omega \leq \pi \tag{5-7a}$$

where

$$\arg_u |X(e^{j0})| = 0 \tag{5-7b}$$

and where the unwrapped-phase derivative is given as in equation (3-6):

$$\arg_u' |X(e^{j\omega})| = \frac{X_R(e^{j\omega})X_I'(e^{j\omega}) - X_I(e^{j\omega})X_R'(e^{j\omega})}{|X(e^{j\omega})|^2} \tag{5-8a}$$

and

$$X'(e^{j\omega}) = X_R'(e^{j\omega}) + jX_I'(e^{j\omega}) = -j\,\mathrm{FT}[nx[n]] \tag{5-8b}$$

Such a phase function is commonly called the *unwrapped phase* of $x[n]$.

As shown in equation (3-12b), the signal lag r_x is simply the mean of the unwrapped-phase derivative. Thus, in general, the unwrapped phase will exhibit a linear-phase component due to r_x which may then be computed as

$$r_x = \frac{\arg_u |X(e^{j\pi})|}{\pi} \tag{5-9}$$

The continuous phase of the normalized input then satisfies

$$\arg |s_x e^{j\omega r_x} X(e^{j\omega})| = \arg_u |X(e^{j\omega})| - \omega r_x \tag{5-10a}$$

In spite of the fact that the unwrapped phase is very precisely defined by equations (5-7) and (5-8), one soon realizes that such equations can never be exactly implemented in a digital computer. It is therefore of interest to investigate alternative methods of evaluating the unwrapped phase. Such methods have been referred to as *phase-unwrapping methods*. Perhaps the most straightforward approach to phase unwrapping is to do numerical integration on the phase derivative, since, through the use of the FFT, it is possible to evaluate very efficiently $\arg_u' |X(e^{j2\pi k/N})|$, $k = 0, 1, \ldots, N-1$, by computing the FFT's of $x[n]$ and $nx[n]$ and combining the results according to equation (5-8).

Another approach that has been taken quite often is to compute the principal value of the unwrapped phase:

$$\mathrm{ARG}_u \,|\, X(e^{j(2\pi k/N)}) \,| = \{\arg_u |\, X(e^{j2\pi k/N}) \,|\}_{\text{mod } 2\pi},$$
$$k = 0, \ldots, M-1 \qquad \text{(5-10b)}$$

using inverse tangent routines on $X(e^{j\omega})$, and then unwrap by appropriately adding multiples of 2π to the principal value until the discontinuities induced by the modulo 2π operation are removed.

Unfortunately, both methods described above depend very critically on the frequency sampling interval $2\pi/N$.

In practice, it is not possible to estimate a priori how large the FFT size N should be in order to guarantee the reliability of these algorithms.

As has been shown elsewhere [13], both methods often lead to very significant unwrapping errors, even for reasonably large FFT's.

V.4 PHASE UNWRAPPING BY ADAPTIVE INTEGRATION

Phase unwrapping has been thoroughly investigated in this book. As a result, a new phase-unwrapping technique has been proposed [13] that combines the information in both the phase derivative and the principal value of the phase into an adaptive numerical integration unwrapping scheme. This new algorithm has proved to be very reliable, and it can be easily matched to the DFT implementation of homomorphic filters.

Let Ω_1 be an arbitrary frequency value and $\mathrm{ARG}_u\,[X(e^{j\Omega_1})]$ be the principal value of the phase at Ω_1. The set of permissible phase values at Ω_1 is then given by

$$\{\mathrm{ARG}_u\,[X(e^{j\Omega_1})] + 2\pi l, \qquad \forall\, l, \text{ integer}\} \qquad \text{(5-11)}$$

The phase-unwrapping problem amounts to determining the correct integer value $l_c(\Omega_1)$ such that

$$\arg_u\,[X(e^{j\Omega_1})] = \mathrm{ARG}_u\,[X(e^{j\Omega_1})] + 2\pi l_c(\Omega_1) \qquad \text{(5-12)}$$

This is done through the use of numerical integration of the phase derivative. We shall adopt here the trapezoidal integration rule. Assuming the unwrapped phase to be known at a frequency $\Omega_0 < \Omega_1$, we shall define a phase estimate at Ω_1, $\tilde{\arg}_u [X(e^{j\Omega_1}) | \Omega_0]$, by

$$\tilde{\arg}_u [X(e^{j\Omega_1}) | \Omega_0] = \arg_u [X(e^{j\Omega_0})] + \frac{\Omega_1 - \Omega_0}{2}[\arg'_u [X(e^{j\Omega_0})] + \arg'_u [X(e^{j\Omega_1})]] \tag{5-13}$$

Clearly, this estimate improves as the step interval $\Delta\Omega = \Omega_1 - \Omega_0$ becomes smaller. We shall define the phase estimate at Ω_1 to be *consistent* if it lies within a predefined distance of one of the permissible phase values at Ω_1, that is, if, for some integer $l_c(\Omega_1)$:

$$|\tilde{\arg}_u [X(e^{j\Omega_1}) | \Omega_0] - \text{ARG}_u [X(e^{j\Omega_1})] + 2\pi l_c(\Omega_1)| < \text{THLD} < \pi \tag{5-14}$$

The basic idea of this algorithm is thus to adapt the step size $\Delta\Omega$ until a consistent phase estimate is found. The resultant $l_c(\Omega_1)$ is used to form the unwrapped phase at Ω_1. This unwrapped phase is then used to form $\tilde{\arg}_u [X(e^{j\Omega_2}) | \Omega_1]$, $\Omega_2 > \Omega_1$, and so on.

The basic idea described previously requires the computation of the phase derivative and of the principal value of the phase on a set of nonequally spaced frequencies. For this algorithm to be practical, one must take full advantage of the FFT algorithm and reduce the number of extra DFT computations to a reasonably small number. Let us denote by $\{\omega_k = (2\pi/N)k,\ k = 0, \ldots, M - 1\}$ the set of uniformly spaced frequencies with interval $2\pi/N$, where $N = 2^M$ or in general any highly composite number. The phase derivative and the principal value of the phase at these frequencies may then be computed using FFT's to evaluate the DFT's of $x(n)$ and $nx(n)$. At each ω_k, a phase estimate is initially formed by one-step trapezoidal integration, starting at ω_{k-1}. If the resultant estimate is not consistent, the adaptive integration scheme is applied within the interval $[\omega_{k-1}, \omega_k]$. The step-size adaptation was carefully designed to minimize the number of extra

DFT's required. The search for consistency is done by consecutively splitting the step interval in half. As the required phase derivatives and principal values are computed, they are stored in a stack fashion. As soon as a consistent estimate is found, the corresponding data are moved out of the stack to a register that holds the most recent consistent estimate of the phase at some frequency within $[\omega_{k-1}, \omega_k]$. New estimates are always formed by integrating from the most recent estimate to the frequency corresponding to the top of the stack. A Fortran program along these lines can be found in [13].

The efficiency of phase unwrapping by adaptive integration can probably be increased, while retaining the reliability of the present algorithm. One possibility is to use splines [15] to match the phase derivative, thus greatly reducing the stepwise numerical integration error associated with the use of the trapezoidal method. This approach is currently being explored [16].

V.5 BANDPASS MAPPING

We shall now propose an efficient implementation for the bandpass-mapping system BP and $(\mathrm{BP})^{-1}$. The implementation of BP is essentially based on a very simple sampling-rate reduction scheme and may be rigorously derived as follows.

Consider the frequency scaling associated with the bandpass-mapping BP:

$$\tilde{\omega} = \pi \frac{\omega - \omega_1}{\omega_2 - \omega_1}, \qquad 0 \leq |\tilde{\omega}| \leq \pi \tag{5-15}$$

Defining the auxiliary frequency mappings

$$\eta(\omega) = \pi \frac{\omega}{\omega_2}, \qquad 0 \leq \eta \leq \pi \tag{5-16a}$$

$$\theta(\eta) = \pi - \eta, \qquad 0 \leq \theta \leq \pi \tag{5-16b}$$

$$\gamma(\theta) = \pi \frac{\theta}{\theta_1}, \qquad 0 \leq \gamma \leq \pi \tag{5-16c}$$

$$\delta(\gamma) = \pi - \gamma, \qquad 0 \leq \delta \leq \pi \tag{5-16d}$$

where $\theta_1 = \theta(\eta_1)$ and $\eta_1 = \eta(\omega_1)$, it can be shown that

$$\tilde{\omega}(\omega) = \delta[\gamma\{\theta[\eta(\omega)]\}], \qquad 0 \leq \tilde{\omega} \leq \pi \tag{5-17}$$

Thus, BP can be realized as the cascade illustrated in Figure 5.2, where each system S_i is associated with a given frequency transformation. Both systems S_2 and S_4 are simple modulators by $\exp[j\pi n]$; that is, their outputs equal their inputs times $(-1)^n$. Systems S_1 and S_3 are digital decimators, the first with a sampling-rate reduction of ω_2/π, the second with a sampling-rate reduction of θ_1/π.

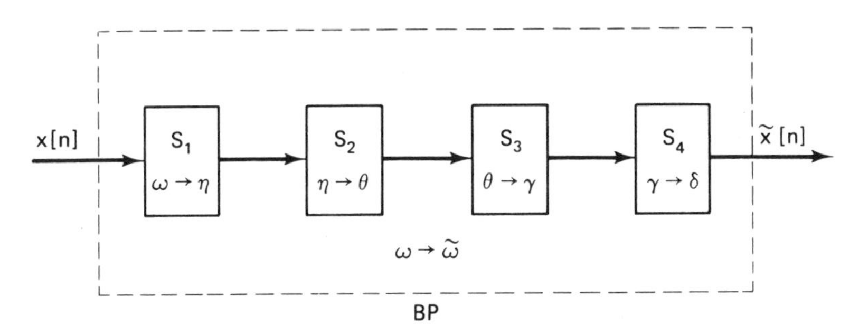

Figure 5.2 Representation of the BP system

Assuming, without loss of generality, that

$$\omega_1 = \omega_2 \cdot \left[1 - \frac{I_1}{D_1}\right] \quad \text{and} \quad \omega_2 = \left(\frac{I_2}{D_2}\right)\pi \tag{5-18}$$

where I_1, I_2, D_1, and D_2 are positive integers, an implementation of BP using single-stage interpolator–decimator structures for S_1 and S_3 is depicted in Figure 5.3. The use of FIR linear-phase low-pass digital filters in the figure is required because the signal characteristics, both magnitude and phase, within the passband (ω_1, ω_2) must be preserved.

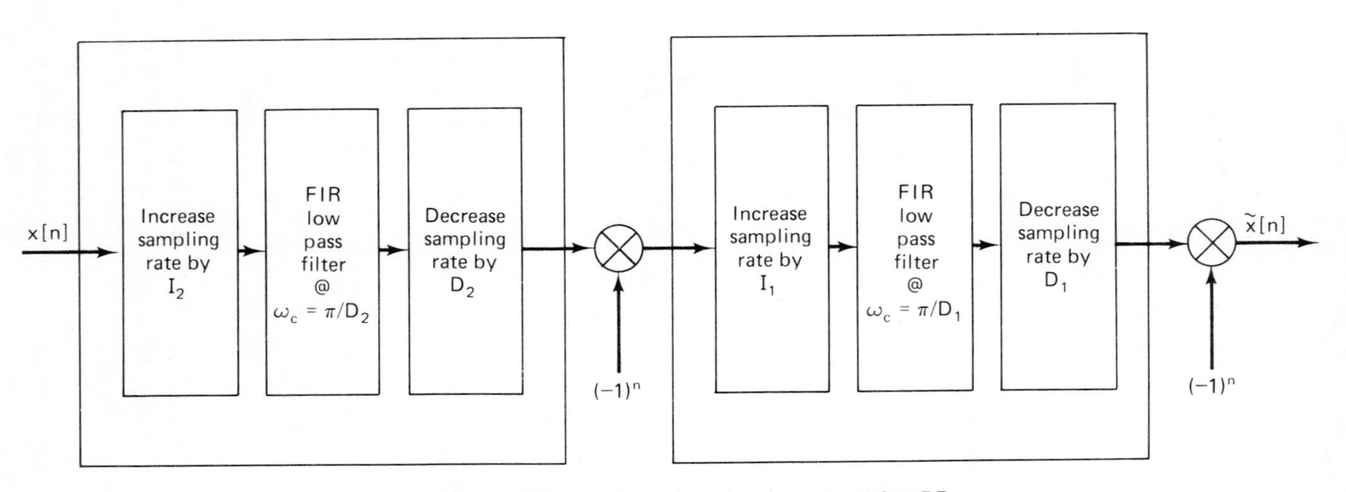

Figure 5.3 Implementation of the bandpass-mapping BP

The design of optimum FIR linear-phase filters for decimation and interpolation has received considerable attention in the literature, and very efficient design algorithms and multistage implementation strategies are now available [14].

When processing large batches of data, it will be necessary to adapt the bandpass-mapping BP to changes in the signal-band cutoff frequencies ω_1 and ω_2. This will require not only changes in the integer sampling-rate increment and decrement factors I_1, I_2, D_1, and D_2 but also changes in the cutoff frequencies of the low-pass filters. Rather than redesigning new FIR linear-phase filters, using the Remez exchange algorithm [17], one might make use of the variable cutoff linear-phase filter structures proposed by Oppenheim, Mecklenbräuker, and Mersereau [18].

The inverse bandpass mapping can be implemented by noting that the frequency mappings of equations (5-16) are invertible within the frequency bands of interest. We conclude then that $(\mathrm{BP})^{-1}$ can be realized as the cascade of S_4^{-1}, S_3^{-1}, S_2^{-1}, and S_1^{-1}, where S_1^{-1} and S_3^{-1} are now digital interpolators.

Finally, we note that the inputs to the bandpass system BP, implemented as described here, do not need to be prefiltered by a linear bandpass filter, since such an operation is automatically performed by the low-pass filters in Figure 5.3. Thus, we shall henceforth denote by $\tilde{r}[n]$ the response of BP to a signal $r[n]$, even though $r[n]$ might not be a bandpass signal in S_B.

Let us now illustrate this approach by bandpass mapping and inverse bandpass mapping the bandpass signal $s[n]$ of Figure 5.4a. The signal satisfies

$$s[n] = w[n] \star r[n] \tag{5-19}$$

where $w[n]$ is the bandpass wavelet of Figure 5.5a and

$$r[n] = \sum_{k=1}^{M} r_k \delta[n - n_k]$$

is the impulse train of Figure 5.6a. The passband cutoff frequencies ω_1 and ω_2 of $s[n]$ and $w[n]$ are given by

$$\begin{aligned} \omega_2 &= \tfrac{1}{2} \cdot \pi \\ \omega_1 &= (1 - \tfrac{5}{6})\omega_2 \end{aligned} \tag{5-20}$$

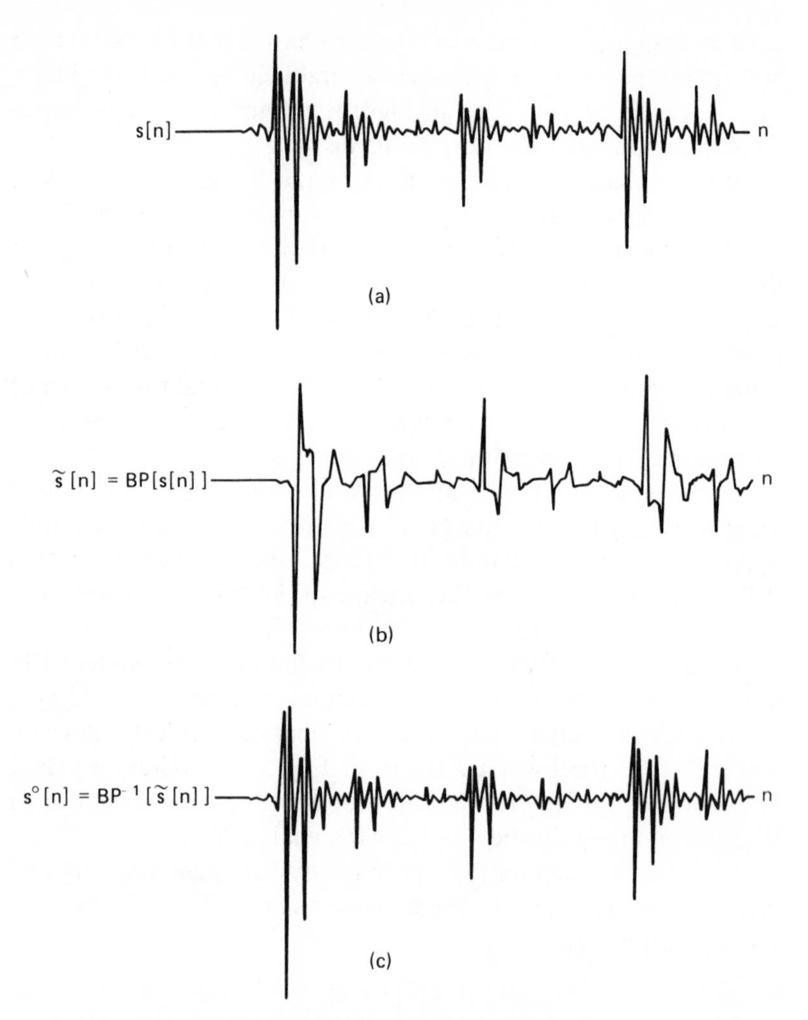

Figure 5.4 Effects of bandpass-mapping a signal $s[n]$ equal to the convolution of a wavelet $w[n]$ with an impulse train $r[n]$

which, assuming a sampling interval of 4 ms, corresponds to a passband from 10.416161 Hz up to 62.5 Hz. Thus, the cutoff frequencies chosen are representative of those one is likely to encounter in seismic reflection studies. The bandpass-mapped components $\tilde{s}[n]$, $\tilde{w}[n]$, and $\tilde{r}[n]$ are illustrated in Figures 5.4b, 5.5b, and 5.6b, respectively. These signals were computed, using the implementation of Figure 5.3, with $I_1 = 5$, $D_1 = 6$, $I_2 = 1$,

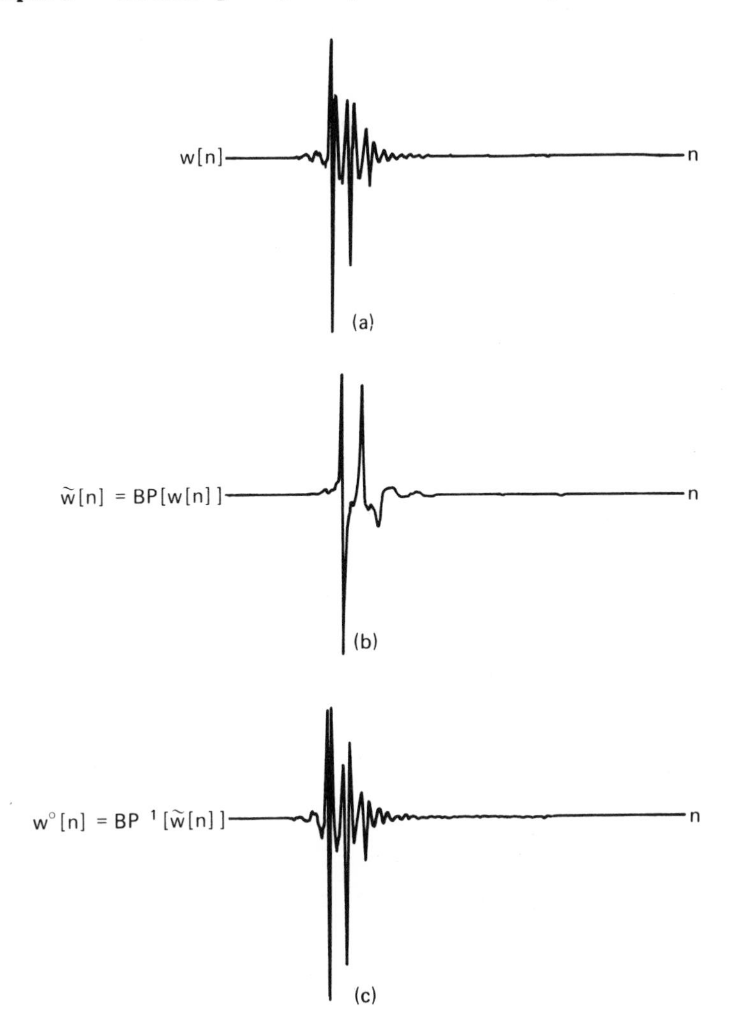

Figure 5.5 Effects of bandpass-mapping a bandpass-filtered wavelet $w[n]$

and $D_2 = 2$, where the ideal low-pass filters called for in the interpolator–decimator structures were approximated by FIR zero-phase optimum Chebyshev filters.

The homomorphic nature of this transformation, discussed in Chapter IV, was verified by convolving $\tilde{r}[n]$ with $\tilde{w}[n]$ and comparing the result with $\tilde{s}[n]$, as depicted in Figure 5.7. The minor discrepancies that exist between the two sequences may be

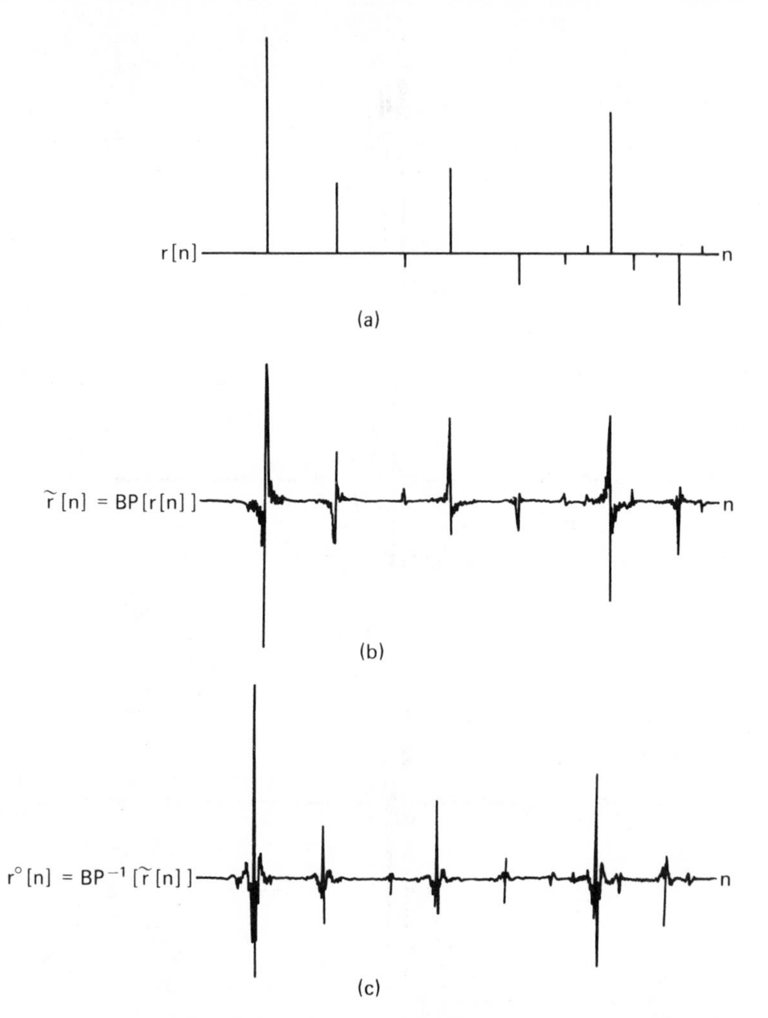

Figure 5.6 Effects of bandpass-mapping a full-band impulse train $r[n]$

rendered negligible by imposing tighter specifications on the FIR filter characteristics.

The structure of $\tilde{r}[n]$ is in good agreement with the theoretical analysis of the bandpass-mapping system BP, as discussed in Chapter IV. That is, we may approximately write

$$\tilde{r}[n] \simeq \tilde{b}[n] \star \tilde{p}[n] \tag{5-21}$$

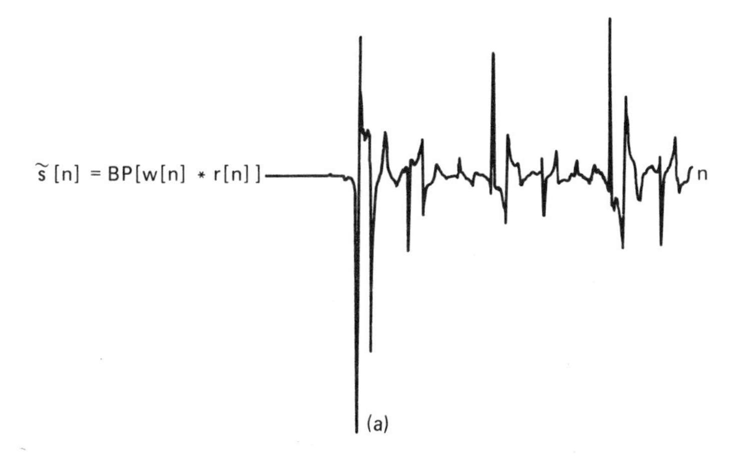

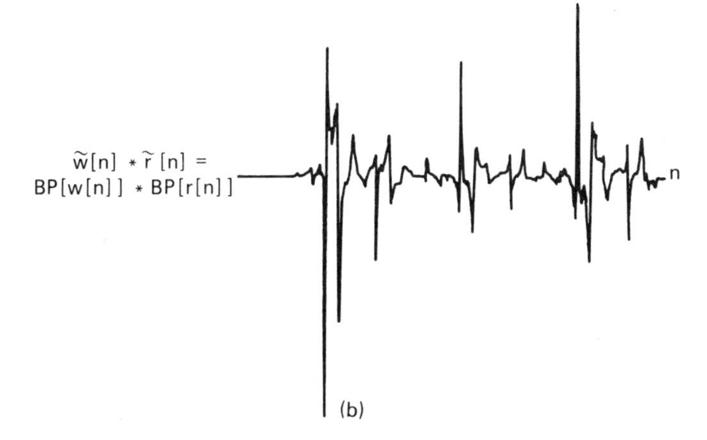

Figure 5.7 Homomorphic nature of the bandpass-mapping BP

where $\tilde{p}[n]$ represents an impulse train of the form

$$\tilde{p}[n] = \sum_{k=1}^{M} \tilde{p}_k \delta[n - \tilde{l}_k] \tag{5-22}$$

and $\tilde{b}[n]$ represents a doubletlike envelope.

The arrival times and amplitudes $(\tilde{l}_k, \tilde{p}_k)$ are related to the arrival times and amplitudes of $r[n]$ as

$$\tilde{l}_k \simeq n_k \frac{I_1 \cdot I_2}{D_1 \cdot D_2}, \qquad k = 1, \ldots, M \tag{5-23}$$

and

$$|r_k| \simeq |\tilde{p}_k|, \qquad k = 1, \ldots, M \tag{5-24}$$

Note, however, that although the basic amplitude structure of $r[n]$ has been preserved in $\tilde{r}[n]$, in the sense of equation (5-24), its polarity has not. This is, in fact, the major time-varying effect that distinguishes a simple sampling-rate reduction ($\omega_1 = 0$) from the more general bandpass mapping and is essentially due to the fact that for $\omega_1 \neq 0$, the cosine amplitude modulation in equation (4-15) is no longer synchronous with the sin x/x function. Thus, the polarity of $\tilde{b}[n, k]$ depends on the instant n_k at which the impulse input was applied to BP. Note, then, that the bandpass mapping essentially preserves the structure of signals such as $s[n]$ in equation (5-19) in the sense that $\tilde{s}[n]$ can still be represented as a convolution of a wavelet $\tilde{w}[n] \star \tilde{b}[n]$ with an impulse train $\tilde{p}[n]$, since, using equation (5-21),

$$\tilde{s}[n] = \tilde{w}[n] \star \tilde{r}[n] \simeq (\tilde{w}[n] \star \tilde{b}[n]) \star \tilde{p}[n] \tag{5-25}$$

Finally, Figures 5.4c, 5.5c, and 5.6c depict the outputs $s^o[n]$, $w^o[n]$, and $r^o[n]$ of the inverse bandpass mapping $(\text{BP})^{-1}$, when the inputs are, respectively, the full-band signals $\tilde{s}[n]$, $\tilde{w}[n]$, and $\tilde{r}[n]$.

As expected, $s^o[n]$ and $w^o[n]$ are essentially identical to $s[n]$ and $w[n]$, with minor discrepancies that can be rendered negligible by using better FIR low-pass filters. The output $r^o[n]$ is essentially identical to the bandpass filtering of $r[n]$.

V.6 SUMMARY

Homomorphic systems may be effectively and reliably implemented using FFT algorithms in conjunction with a new phase-unwrapping technique, based on an adaptive numerical integration scheme. The input normalization is automatically done at the phase-unwrapping stage.

The bandpass mapping may be implemented using simple interpolator–decimator structures. The inputs to the system do not need to be prefiltered by a bandpass filter, since such an operation is automatically incorporated in this implementation.

VI

Seismic Modeling

VI.1 INTRODUCTION

The present chapter is dedicated to the study of seismic signal models, both in exploration and in teleseismic situations, with the purpose of clearly identifying the various seismogram components as well as defining the basic goals of seismic data analysis.

VI.2 SEISMIC REFLECTION MODELS

Among the geophysical prospecting techniques, reflection seismology is by far the most widely used. With this method the subsurface structure is mapped by measuring the times required for a seismic wave, generated in the earth by the excitation of a near-surface acoustic source, to return to the surface after reflection from interfaces between formations having different acoustic impedances.

The reflected waves are recorded at the surface by sensors whose outputs describe the variation in time of pertinent physical variables. Such recordings, commonly called seismograms or seismic traces, are subject to a number of signal-processing operations, followed by interpretation in terms of depths and reflectivities of each of the subsurface reflecting horizons. Such interpretation is an extremely complex task that requires familiarity with the basic physical principles governing the propagation characteristics of seismic waves within the earth. These include their generation, transmission, absorption, and attenuation in the earth materials, as well as their reflection, refraction, and diffraction characteristics at discontinuities.

Seismic wave propagation is now quite well understood; good texts at the introductory and advanced level are, respectively, Dobrin [19] and Grant and West [20].

Earth Models

As far as seismic waves are concerned, the earth behaves as a linear medium. Thus, one may characterize its behavior in terms of its response to an impulsive seismic source, as illustrated in Figure 6.1. Such characterization is often done with the help of simplified earth models, whose purpose is to facilitate the understanding of the generation of the various seismogram components as well as to capture the basic phenomena that motivate specific signal-processing tasks.

In general, the signal arriving at each surface receiver is a complex of a number of distinct components, some of them to be considered as signal since they carry information that enables the potential recovery of the subsurface information we are looking for; some of them to be regarded as noise, since they either have no relationship with the subsurface structure or such a relationship is too complex to be considered in a processing scheme.

Conventionally, the signal is defined to be the reflective component in a seismogram, everything else being noise. A primary signal-processing task in reflection seismology is thus the elimina-

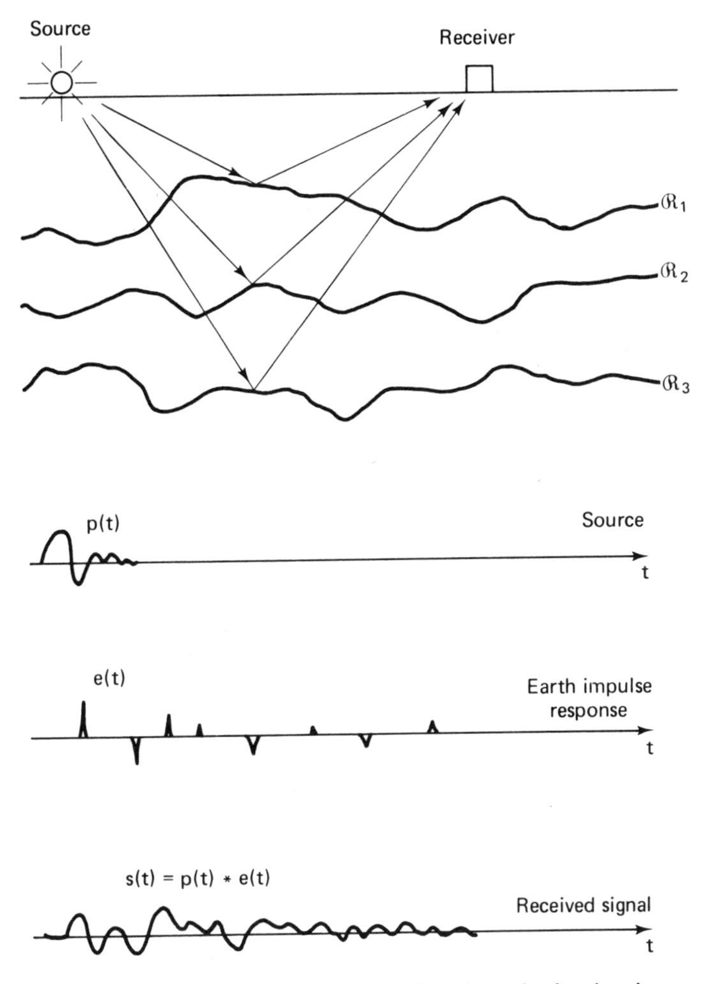

Figure 6.1 Simplified model representing the seismic signal as a convolution of components

tion of such unwanted noise components; fortunately, signal and noise waves propagate with different speeds and in different directions in the earth, thus allowing eventual discrimination by means of appropriate source and receiver array geometries in conjunction with powerful multichannel signal-processing techniques, for example, CDP stacking [21].

We shall thus assume throughout this book that the seismograms we need to be concerned with have already been stripped of most noisy components in the sense mentioned above. We shall then attempt to describe, by means of the corresponding earth models, the generation of reflective components within the earth. Such simplified models were introduced as early as 1958 by Berryman et al. [22] and were successively developed by Wuenschel [23], Trorey [24], Claerbout [25], Middelton and Whittlesey [26], Treitel and Robinson [27], and others. The feature common to all seismic models is that the earth consists of a half-space of infinite parallel-plane homogeneous layers. We shall describe next a model recently introduced by Theriault [28]; its main assumptions can be summarized as follows:

1. All waves are plane, longitudinal compressional waves, traveling normal to the layer interfaces.
2. Each layer is characterized by a thickness x_i, a velocity c_i, and an attenuation parameter α_i.
3. The ith interface separates the ith and the $(i+1)$st layers and is characterized by the reflection coefficient r_i, which is given by

$$r_i = \frac{c_{i+1} - c_i}{c_{i+1} + c_i} \tag{6-1}$$

   where the layer densities are assumed to be equal to each other.
4. Each layer attenuates waves passing through it by the same mechanism, although the degree of attenuation may vary from layer to layer.

The attenuation of elastic waves in rocks is a well-established phenomena, although the physical attenuation mechanisms are still not well known. Attenuation characteristics as observed in the laboratory or in the field for representative rock samples have been extensively tabulated. The departure of Theriault's model from conventional models lies essentially in the proposed transfer function thought to represent the lumped attenuation and travel delay undergone by an impulsive wave during one passage through one earth layer. For the ith earth layer, such a transfer function is

$$F(\omega, \alpha_i, \tau_i) = (j\omega\alpha_i)^{-3} \cdot \exp(-j\omega\tau_i) \tag{6-2}$$

where the travel delay $\tau_i = x_i/c_i$. The proposed attenuation law manifests an accurate adherence to the qualitative properties observed experimentally, over the seismic frequency range of interest, 10–250 Hz.

The lumped-parameter model of a single-layer interface is illustrated in Figure 6.2, where $U_i(\omega)$ and $D_i(\omega)$ are the Fourier

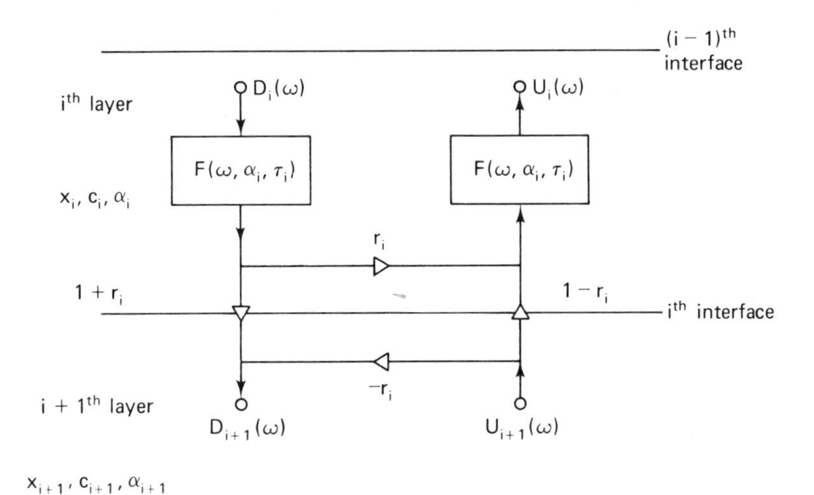

Figure 6.2 Lumped-parameter model of single-layer interface

transforms of the pressures associated with the upward and downward traveling waves, respectively, as observed at a point just below the $(i-1)$st interface. The four gains, r_i, $-r_i$, $1 + r_i$, and $1 - r_i$, model the reflective and transmissive phenomena at the ith interface. As illustrated in Figure 6.3, a lumped-parameter model for the entire earth can thus be formed by concatenating a number of single-layer interface models. Since the earth is a distributed-parameter system, such concatenation of a number of

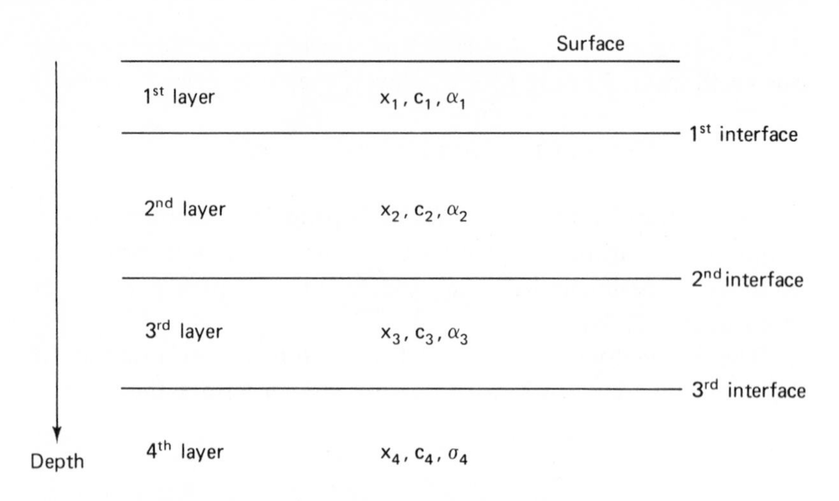

Figure 6.3 Plane-layered homogeneous earth model

lumped-parameter transfer functions must obey the following semigroup property:

$$F(\omega, \alpha_i, \tau_i) \mathbf{O} F(\omega, \alpha_j, \tau_j) = F(\omega, \alpha_i + \alpha_j, \tau_i + \tau_j) \tag{6-3}$$

where we define the operator **O** to represent the concatenation of layers i and j into one single lumped-transfer function.

Using Theriault's earth model, it can be easily shown that the overall earth transfer function can be simply computed by means of parametric manipulation of amplitudes, attenuations, and travel times. The earth impulse response $e(t)$ can then be written in terms of an infinite sum of scaled, delayed characteristic functions $f(t, \gamma)$:

$$e(t) = \sum_{n=1}^{\infty} A_n f(t - T_n, \gamma_n) \tag{6-4}$$

where

$$f(t, \gamma) = \frac{t^2}{2\gamma^3} \cdot \exp\left(\frac{-t}{\gamma}\right) u_{-1}(t) \tag{6-5}$$

Equation (6-4) expresses mathematically what is commonly called *Ricker's wavelet theory of seismogram structure* [29]: each seismogram consists of many wavelets with different strengths and arrival times, due to disturbances that have traveled different source-to-receiver paths and have suffered different degrees of attenuation.

The response $s(t)$ of a plane-layered earth structure with an impulse response $e(t)$ to a seismic plane compressional transient wave $p(t)$, traveling normal to the layer interfaces is then, by linearity,

$$s(t) = p(t) \star e(t) = \sum_{n=1} A_n w(t - T_n, \gamma_n) \qquad (6\text{-}6a)$$

where

$$w(t, \gamma) = p(t) \star f(t, \gamma) \qquad (6\text{-}6b)$$

is now the basic seismic wavelet.

Near-Surface Effects

So far we have characterized the response of an idealized earth structure to an idealized seismic excitation. In practice, the seismic source is either buried near the surface, as happens with dynamite sources; at the surface, as happens with Vibroseis; or within the water layer, as in marine profiling.

The near-surface structure and geometry often impose specific constraints on the illumination of the deeper reflectors. In land, the upper sediments, being normally not fully consolidated—the *weathered layer*—affect the propagating seismic pulse in ways that cannot be simply modeled, but rather described in terms of an overall delay and significant distortion of the source pulse as it travels downward. The net result of this weathered layer is that the transient wavelet arriving at the deep reflectors no longer resembles the shot pulse. Of course, the weathered layer will also affect the upcoming energy. Thus, although the general wavelet seismic structure of equation (6-6a) may still be valid, the relationship between each wavelet and the shot pulse is now much more complicated then the one in equation (6-6b). As a consequence, the knowledge of the source pulse is of little use in the eventual recovery of the wavelet arrival times, in those situations where weathered-layer effects are significant.

Reverberation

Another example where the near-surface structure strongly affects the illumination of the deeper structure is when the source is buried beneath a strong reflector. Here, the source energy that traveled upward and was reflected downward by such reflector creates a secondary illumination that generates ghosts of each deep reflector. This is, of course, always the case in marine shooting, since the sea–air interface acts like a perfect reflector at seismic frequencies. Furthermore, whenever the sea bottom or the near-subsurface reflectors are good reflectors, a reverberating effect will occur as a result of the trapping of most source energy in the water layer. In such a case, the deeper reflections may be overlapped by the arrivals of these water-trapped multiple reflections, leading to significant data-analysis problems.

Water-column reverberation may often be modeled in terms of simple parametric models, from which specific data-processing techniques are designed, with the basic goal of enhancing the deep primary reflections over the water-column multiples. The key to such models is the periodic nature of the reverberation phenomena. As illustrated in Figure 6.4, the reverberation of the source energy in the water layer, the *first-order ringing*, can be modeled as the convolution of the source pulse $p(t)$ with a periodic train $m_1(t)$ of the form

$$m_1(t) = \sum_{k=0}^{\infty} (-1)^k r_1^k \delta[t - kT_w] \tag{6-7}$$

where r_1 denotes the reflection coefficient of the sea bottom and T_w the two-way travel time through the water layer.

This same reverberation mechanism also acts upon the upcoming reflected energy, the *second-order ringing*, leading thus to models of the seismic trace $t(t)$ as

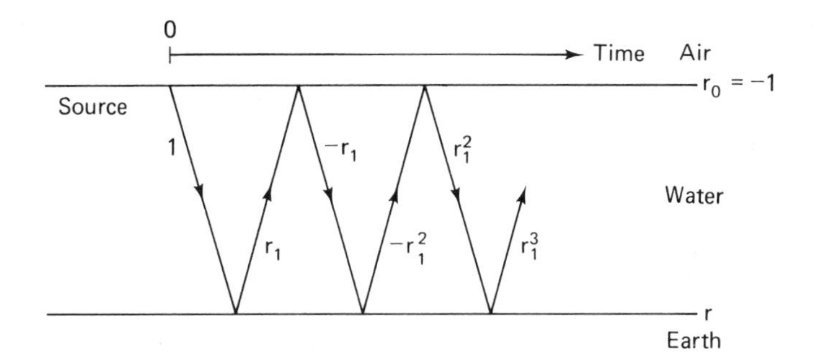

Figure 6.4 Source reverberation mechanism

$$\begin{aligned} t(t) &= \{[p(t) \star m_1(t)] \star e(t)\} \star m_1(t) \\ &= s(t) \star m(t) \end{aligned} \tag{6-8}$$

where

$$m(t) = m_1(t) \star m_1(t) \tag{6-9}$$

denotes the complete reverberation operator (first- and second-order ringing).

The removal of seismic reverberations, which we shall henceforth denote as *seismic dereverberation*, has been extensively treated in the literature. A number of filtering techniques have been developed for this purpose, the most common of which are undoubtedly those based on predictive filtering schemes [30].

Additive Noise

Reflection data will, of course, always deviate from the idealized representations described above, owing to the presence of noise components that happened to be traveling with speeds and in a direction similar to the reflected energy.

Fortunately, this noise, which originates from a variety of sources (e.g., ambient, microseismic, etc.) is often concentrated at low or at high frequencies. It is thus a standard practice to bandpass-filter the seismograms in order to improve the reflection signal-to-noise ratio.

Signal Analysis

As pointed out earlier, the basic goal of seismic signal processing consists in the determination of wavelet arrival times and amplitudes.

Currently, signal processing is mostly carried off-line, on digital-computer facilities. Seismic data, being limited to frequencies below, say, 250 Hz, because of the attenuation characteristics of the earth, is low-pass-filtered, sampled, and digitally recorded on-line.

Let us represent the basic wavelet sequence as $w[n, \alpha]$ and let us represent the wavelet arrival times as $(n_i, i = 0, 1, \ldots)$. The seismic trace $t[n]$ can then be modeled as

$$t[n] = s[n] \star m[n] + \eta[n] \tag{6-10}$$

where $s[n]$ represents the response of a plane-layered earth structure to a seismic-plane compressional wave, of the form

$$s[n] = \sum_{k=1}^{\infty} A_k w[n - n_k, \gamma_k] = p[n] \star e[n] \tag{6-11}$$

where $m[n]$ denotes an operator that represents near-surface multiple generation (e.g., a reverberation operator of the form

$$m[n] = \sum_{k=0}^{\infty} m[k]\delta[n - kTw] \tag{6-12}$$

which might be assumed to be an impulse $\delta[n]$, when such effects are negligible), and where $\eta[n]$ represents an additive noise component.

Ideally, then, the seismic signal processor should be able to eliminate the noise and the reverberation operator and to compress each wavelet $w[n - n_k, \gamma_k]$ into an impulse $\delta[n - n_k]$, as illustrated in Figure 6.5.

We shall denote the output $r[n]$ of such an ideal processor as the *seismic reflector series;* this would be the response of an ideal lossless (nondispersive) earth to an impulsive source.

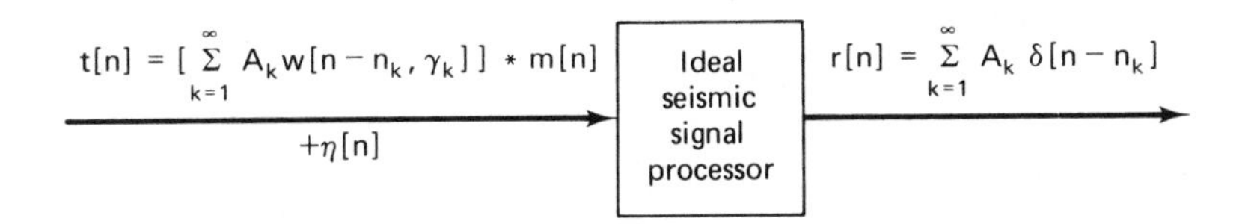

Figure 6.5 Ideal seismic signal processor

Such an ideal compressing scheme is clearly impossible to design in practice. One must thus seek not only more appropriate data modeling schemes but also processing algorithms which, being bound to deviate from the ideal performance, are robust under a wide range of noise conditions and yield, in practice, sharp, short bursts of energy at each of the arrival times n_k.

VI.3 TELESEISMIC MODELS

Seismic methods are also widely used in the study of earthquakes and explosive phenomena within the earth. Here, the source of seismic energy is buried within the earth, often at great depths, and waves are transmitted through the earth and at its surface, across great distances.

Often it is of interest to analyze the compressional component (P-wave) generated by such phenomena. P-waves are transmitted through the earth by a mechanism similar to that previously described. Thus such seismograms can be modeled as the convolution of the source pulse $p[n]$ with the impulse response of the transmission path $e[n]$. The transmission effects through the earth can be essentially divided into three categories: (1) near-source effects, (2) transmission through the mantle, and (3) near-receiver effects.

The transmission through the mantle greatly attenuates the seismic pulse, which is now restricted to very low frequencies. At these frequencies, the earth layers underneath the recording site are basically nondispersive. Thus, teleseismic data models are often of the form

$$s[n] = w[n] \star r[n] + \eta[n]$$

where $w[n]$ represents the seismic pulse as it emerges from the mantle, $r[n]$ represents the transmission impulse response of the earth structure at the recording site, and $\eta[n]$ represents a noise component. The recovery of the seismic pulse $w[n]$ is of great importance in the study of elastic-wave attenuation and dispersion in the earth and in the understanding of earthquake mechanisms. Thus, the major goal in teleseismic signal analysis is the recovery of the seismic wavelet $w[n]$.

VI.4 SUMMARY

Reflection seismograms are essentially composed of many wavelets with different strengths and arrival times. These wavelets are, in general, distinct from one another in that they have suffered different attenuations while traveling through the earth layers.

The major goal in the analysis of reflection seismograms is the recovery of the earth reflector series. This ideally implies the removal of any near-surface multiple mechanisms, which might create ghosts or reverberations, as well as the contraction of each individual wavelet into an impulse.

Teleseismic data are often modeled as the convolution of a seismic wavelet with an impulse train representing the transmission impulse response of the earth at the recording site. The major goal in teleseismic data analysis is the recovery of the seismic wavelet.

VII

Seismic Deconvolution

VII.1 INTRODUCTION

Most seismic analysis schemes in use today are based on representations of a seismic trace $s[n]$ as a convolution of two components:

$$s[n] = w[n] \star r[n] \tag{7-1}$$

where $w[n]$ denotes the seismic wavelet and $r[n]$ the seismic reflector series.

Such representation may be derived from the model of equation (6-6a), by assuming that the earth attenuation effects are negligible, so that all wavelets present in a seismic trace are essentially alike.

Denoting the common shape of these wavelets as $w[n]$, it follows that

$$s[n] = \sum_{k=1}^{\infty} A_k w[n - n_k] = w[n] \star \left[\sum_{k=1}^{\infty} A_k \delta[n - n_k] \right] \qquad (7\text{-}2)$$

which is precisely the model of equation (7-1).

We shall refer to models of this type as *time-invariant seismic models*, to emphasize the stationarity of the wavelet shape in time.

For this class of models, and in the absence of noise, the ideal reflection seismic processor has essentially to contract $w[n]$ to an impulse. In other words, it must act as a deconvolution filter. The use of wavelet contraction schemes to achieve increased arrival times resolution was proposed as early as 1953 by Norman Ricker [31]; based on his theoretical studies on the form and nature of seismic wavelets, Ricker proposed the use of "equalization filters," deterministically designed from the theoretical characteristics of seismic wavelets. At about the same time, the pioneering paper by members of M.I.T.'s Geophysical Analysis Group on statistical seismic data modeling and the use of linear operators was published [32]. These two approaches were to be combined a few years later by Robinson [33] to yield a structural stochastic-process description of seismic traces, which essentially consisted in a representation as in equation (7-1), where the seismic reflector series $r[n]$ was modeled as an uncorrelated Poisson series.

This framework was invaluable for the development of linear seismic analysis techniques, the most notable of which are certainly *predictive deconvolution* and *digital Wiener filtering*.

More recently, the work of Oppenheim and Schafer [4] in speech analysis and echo detection and removal by homomorphic filtering has aroused the interest of the seismic community in such systems. The application of homomorphic filtering to the analysis of time-invariant seismic models was first proposed by Ulrych [34] and later developed by many others.

Both linear and homomorphic methods have peculiar advantages and disadvantages, their performance on real data analysis being dependent on the degree to which the data fit each method's most crucial assumptions.

In this chapter we shall review both linear and homomorphic seismic deconvolution techniques. The review of linear deconvolution techniques in Section VII.2 will be limited to the two methods referred to above: predictive deconvolution and digital Wiener filtering. As we shall show later, these methods are particularly suited for use with homomorphic signal analysis. The review of homomorphic deconvolution of seismic data in Section VII.3 will be fairly extensive and detailed. We shall summarize the various homomorphic analysis strategies discussed in the geophysical literature. These strategies are characterized by the fact that they are based upon the time-invariant seismic model of equation (7-1). This class of techniques will be referred to as time-invariant homomorphic analysis techniques. We shall clearly delineate the major assumptions associated with each strategy, as well as point out the major problems encountered in its application.

VII.2 LINEAR DECONVOLUTION

Consider a seismic trace $s[n]$ as in equation (7-1), and let $r[n]$ be uncorrelated. Denoting by $R_{xx}[n]$ the autocorrelation function (ACF) of a sequence $x[n]$, the preceding assumption implies that

$$R_{rr}[n] = R_{rr}[0]\delta[n] \tag{7-3}$$

Therefore, the ACF of such seismic trace is simply proportional to the ACF of the corresponding wavelet, since

$$R_{ss}[n] = R_{ww}[n] \star R_{rr}[n] = R_{rr}[0]R_{ww}[n] \tag{7-4}$$

Before we present the design of specific deconvolution filters, solely based on the information provided by the trace's ACF, let us investigate its sufficiency in terms of the design of an ideal inverse filter for $w[n]$. Let $f_I[n]$ be such an ideal filter. Letting $F_I(e^{j\omega})$ denote its Fourier transform, it follows that

$$F_I(e^{j\omega}) = W(e^{j\omega})^{-1} \tag{7-5a}$$

or, equivalently,

$$|F_I(e^{j\omega})| = |W(e^{j\omega})|^{-1} \tag{7-5b}$$

$$\arg|F_I(e^{j\omega})| = -\arg|W(e^{j\omega})| \tag{7-5c}$$

This ideal inverse filter may then be modeled as a cascade of a zero-phase filter with transfer function given by equation (7-5b), with an all-pass filter with transfer function given by equation (7-5c). Obviously, the trace's ACF conveys enough information to design the first of these systems, but in general provides no information to design the other. An important exception is when the wavelet $w[n]$ is minimum-phase, since then its phase may be derived from its magnitude by means of a Hilbert transformation [1]. Similar exceptions occur when the wavelet is maximum-phase.

In practice, the magnitude and phase characteristics of a deconvolution filter must be designed from the data, because the wavelet $w[n]$ is, in general, not known. Even if an estimate of $w[n]$ were available, it is usually not desirable to perform deconvolution by means of exact frequency division, as postulated by the ideal filter of equations (7-5), since the model of equation (7-1) must be looked upon as an approximate representation of real seismic data. Thus, deconvolution filters must be designed which approximate the ideal behavior in a controlled way.

Two classes of linear filters that have been used extensively in seismic data analysis are digital Wiener filters [35] and predictive deconvolution filters [30]. Both filters are designed by parametric least-squares methods.

Digital Wiener Filters

Wiener filters are linear time-invariant filters that attempt to optimally transform, in the least-squares sense, a given signal into another. In the context of seismic deconvolution, the goal is, thus, to convert the seismogram $s[n]$ into the reflector series $r[n]$. To make this problem meaningful for calculation on a digital computer, one restricts the filters to be finite impulse response (FIR) filters. The resulting class of FIR least-squares filters has been commonly referred to in geophysics as *digital Wiener filters*.

In defining an FIR filter, one has to consider both its length and its time registration. Let us denote by $f[n; P, L]$ a filter of length P, located within the interval $[-L, P - L + 1]$, that is,

$$f[n; P, L] = f[n; P, L] \cdot [u[n + L] - u[n - P + L]] \tag{7-6}$$

We shall refer to P as the filter's order and to L as the filter's lag. The digital Wiener filter of order P and lag L is then defined in terms of the minimization of the l_2 error norm:

$$E(P, L) \triangleq \sum_{n=-\infty}^{\infty} [s[n] \star f[n; P, L] - r[n]]^2 \tag{7-7}$$

Denoting, in general, the cross-correlation between sequences $x[n]$ and $y[n]$ by $R_{xy}[n]$, that is,

$$R_{xy}[n] = \sum_{k=-\infty}^{\infty} x[n]y[n - k] \tag{7-8}$$

it follows that the filter coefficients $f[n; P, L]$ must satisfy the set of normal equations

$$\sum_{i=-L}^{P-L-1} f[i; P, L]R_{ss}[i - j] = R_{rs}[j], \qquad j = -L, \ldots, P - L - 1 \tag{7-9}$$

These equations can be solved very efficiently using Levinson recursion techniques [36].

Since $r[n]$ is assumed to be uncorrelated, it follows that

$$R_{rs}[j] = R_{rr}[0]w[-j] \tag{7-10}$$

so that the normal equations become

$$\sum_{i=-L}^{P-L-1} f[i; P, L]R_{ww}[i - j] = w[-j], \qquad j = -L, \ldots, P - L - 1 \tag{7-11}$$

It can be verified that these are the normal equations for the digital Wiener filter that optimally transforms the seismic wavelet $w[n]$ into a unit spike $\delta[n]$. For this reason, this class of Wiener filters is commonly referred to as *spiking filters*.

The performance of Wiener filters, measured in terms of the least-squares error norm, will obviously depend not only on the filter length P, but also on its lag L [37]. It is thus important to search for the optimum-lag Wiener spiking filter, or order P, such that

$$E(P, L_{\text{opt}}) \leq E(P, L), \qquad \forall L \tag{7-12}$$

It can be shown that Wiener filters of given length but different delays can be derived from each other by the Simpson sideways recursion technique [36]. Thanks to these recursive algorithms, it is possible to search efficiently for optimum-lag Wiener filters.

The use of Wiener filters in seismic deconvolution has been limited by the fact that their use requires knowledge of the seismic wavelet. Since such knowledge is not readily available from the data and in the absence of reliable signal-processing algorithms that are able to extract it, the approach has been to restrict the filter lags to intervals for which the right-hand side of equation (7-11) vanishes. Letting, without loss of generality, $w[n] = 0$, $n < 0$, we conclude that the zeroth-lag Wiener spiking filter satisfies

$$\sum_{k=1}^{P-1} f[k; P, 0] R_{ww}[k - l] = w[0]\delta[l], \qquad l = 0, \ldots, P - 1 \tag{7-13}$$

which enables the determination of such Wiener filters from ACF data alone, except for a gain factor. These filters were originally developed in terms of predictive filtering theory, as discussed next. Although simple to design, these filters are obviously not capable, in general, of providing proper phase compensation, since the one they provide is appropriate only for minimum-phase wavelet compression [37].

Predictive Deconvolution Filters

In the presence of reverberation, a seismic trace is often modeled as in equation (6-8):

$$t[n] = s[n] \star m[n] = w[n] \star m[n] \star r[n] \tag{7-14}$$

where $m[n]$ is a periodic reverberation operator.

One of the most powerful seismic dereverberation techniques is the method of predictive deconvolution, originally proposed by Robinson [33] and later developed by Peacock and Treitel [30]. In essence, a predictive deconvolution filter $f[n]$ is an FIR filter with transfer function $F(z)$ given by

$$F(z) = 1 - z^{-D}A(z) \tag{7-15}$$

where

$$A(z) = \sum_{k=0}^{P-1} a[k]z^{-k} \tag{7-16}$$

called the *prediction operator* is a zero-lag Wiener filter of order P, which attempts to predict $t[n + D]$, a time-advanced version of the input $t[n]$. The value D is referred to as the prediction distance. The prediction operator thus satisfies the following set of normal equations:

$$\sum_{i=0}^{P-1} a[i]R_{tt}[i - j] = R_{tt}[D + j], \qquad j = 0, \ldots, P - 1 \tag{7-17}$$

If $r[n]$ is assumed to be uncorrelated, these equations become

$$\sum_{i=0}^{P-1} a[i]R_{w_m w_m}[i - j] = R_{w_m w_m}[D + j], \qquad j = 0, \ldots, P - 1 \tag{7-18}$$

where we denote by $w_m[n]$ the reverberated seismic wavelet, that is,

$$w_m[n] = w[n] \star m[n] \tag{7-19}$$

It can be shown that, by appropriately selecting the prediction distance D to be less than or equal to the reverberation period T_w and the length of the prediction operator P to be greater than or equal to T_w, both first- and second-order ringing can be removed by such filters.

Prediction filters are determined in practice by solving equation (7-17). Their ability in effectively removing the source reverberation depends crucially on whether or not such solution is actually close to the solution of equation (7-18). In other words, it depends on whether or not the reflector series is, in fact, uncorrelated up to lag $n = P + D$. As shown by Baggeroer [38], this tends to be the case in deep-water profiling, while it is often not so in shallow water.

It can be easily verified that the predictive filter with unit prediction distance, also referred to as linear prediction filters, are identical to the zero-lag Wiener spiking filters except for a constant gain.

Linear prediction is often used as a basis for the estimation of the seismic wavelet, given the trace's ACF. The z-transform of the wavelet $W(z)$ is estimated from the z-transform of the linear prediction filter as

$$W(z) = \frac{G}{1 - \sum_{k=1}^{P} a_k z^{-k}} \tag{7-20}$$

where G is a constant. As we pointed out before, this representation is only appropriate for minimum-phase wavelets.

VII.3 HOMOMORPHIC DECONVOLUTION

Seismic analysis by homomorphic filtering explores the distinct characteristics of the signal components in the cepstral domain. It is thus a powerful and promising technique in that it does not require a priori knowledge of the shape of the seismic wavelet. Furthermore, the seismic wavelet is not assumed to be minimum-phase, nor is the reflector series assumed to be uncorrelated, as happens with the predictive deconvolution and Wiener filtering techniques.

The homomorphic signal analysis strategy must take into account both the signal models and the analysis goals. The various strategies that have been explored for seismic analysis are characterized by the fact that they are based upon the time-invariant seismic model of equation (7-1). Thus, the complex cepstrum of a seismic trace $\hat{s}[n]$ consists of a component associated with the seismic wavelet plus a component associated with the reflector series.

The spectra of seismic wavelets are, in general, much smoother than that of the reflector series. Thus, $\hat{w}[n]$ will tend to be concentrated around the time origin.

The complex cepstrum of a reflector series, $\hat{r}[n]$, will, in general, have a complicated structure. In order to render this structure as simple and predictable as possible, while contributing to the separation of the signal components in the cepstral domain, it has been a common practice to exponentially weight the data, in order to make the reflector series minimum-phase. Then, as discussed in Chapter III, the reflector series will contribute to the complex cepstrum only for positive times greater than or equal to the interval between the first two arrivals. In such a case, the use of a cepstral window around the time origin, which corresponds to smoothing (i.e., low-pass filtering of the log spectrum) would tend to retain the contributions from the source pulse, whereas a window that retained only the high-time portion of the complex cepstrum, which corresponds to high-time filtering the log spectrum, could be used to recover such minimum-phase reflector series.

The different techniques published in the geophysical literature share the assumption that the reflector series is either minimum-phase or has been made so by exponential weighting. These techniques differ essentially with regard to the particular analysis goals and the procedures by which the recovery of the signal components in the complex cepstrum is accomplished. It should be kept in mind that all these techniques make use of full-band homomorphic systems, even though the data may have bandpass characteristics. Furthermore, phase unwrapping was accomplished by conventional, nonadaptive algorithms.

Teleseismic Analysis

Homomorphic deconvolution of seismic data was first explored by Ulrych [34], and was directed toward the deconvolution of teleseismic events. Ulrych proposed to recover the seismic wavelet by low-time cepstral gating, and presented a number of examples using noiseless as well as noisy seismograms. He observed that if the transmission response was minimum-phase, the low-time portion of the complex cepstrum of a noisy seismogram could still be used to recover the seismic wavelet. However, the estimates of the transmission response by high-time cepstral gating become very noisy, which was interpreted as being due to the irregularities caused in the unwrapped phase by the additive noise. The noise was thought to influence the high-time portion of the complex cepstrum to a much larger degree than it did in the short-time portion. An explanation of Ulrych's findings will be given later.

This technique was applied by Ulrych et al. [39] to deconvolve three earthquakes and one underground nuclear explosion, recorded at Leduc, Alberta. The theoretical impulse response at the recording site was known, so that an estimate of T_1 was available and was used as the cutoff for the low-time gating operation. The quality of the wavelet estimates was verified by convolving it with the theoretical impulse response for the recording site and comparing the resulting synthetic trace with the recorded original. Striking similarities were found between the two.

Clayton and Wiggins [40] considered the problem of deconvolving seismograms for which the seismic wavelets were not confined to the low-time cepstral region, owing to the presence of sharp source antiresonances, corresponding to zeros near the unit circle. Because of such zeros, the complex cepstrum of the source will have a long duration and in particular will tend to overlap with the complex cepstrum of the impulse response.

They next considered a suite of teleseismic recordings of the same event. The suite was restricted in azimuthal and delta ranges so that an assumption of spatial source stationarity could be made. The seismograms correspond thus to the convolution of the same source pulse with different transmission-path impulse responses. The source shape was estimated by averaging the log amplitude and phase spectra of the recordings. This estimation method was thought to be better than simply averaging a particular arrival over several traces, after suitable time alignment, in that it uses the redundant information contained in all arrivals. We shall, however, show that this is not the case, when the impulse response is minimum-phase. It was argued that the necessary conditions for this estimator to resolve the source wavelet is that the various arrivals must have different moveouts along the suite of recordings.

Subsequent deconvolution of the seismograms was done by a modified spectral division scheme, using the wavelet previously estimated. This approach was used to analyze recordings of the Kern County (California) earthquake with good results.

Sommerville [41] addressed the problem of recovering the seismic wavelet by low-time gating, when no a priori information exists regarding the impulse response at the recording site, and, in particular, regarding the time interval between the first two arrivals. Furthermore, he considered that only one recording was available for the analysis. He then devised a scheme by which a series of wavelet estimators were computed using a wide range of cepstral cutoff values. The wavelet chosen was the one that occupied an approximately median position within the envelope of wavelet estimates. Sommerville also observed that the effects of additive noise on the unwrapping of the phase curve, which is very noticeable in the region of low-spectral energy, originated vastly different phase curves, yet caused no difficulties in this procedure, in that the recovered wavelets were very insensitive to these effects. He deconvolved a large number of seismograms, recorded at various stations and originating from two earthquake events, with good results.

Seismic Reflection Analysis

Homomorphic analysis was first applied to seismic reflection data by Stoffa, Buhl, and Bryan [42]. Their work focused on shallow-water marine seismic data, obtained from a free-firing airgun and a single-channel detector streamer.

Homomorphic filtering was used with the dual purpose of source deconvolution and water-column dereverberation. The seismic trace $t[n]$ was modeled as in equation (7-14). For data of this type, the source pulse corresponding to the airgun signature exhibits a bubble-pulse periodicity. The airgun pulse $w[n]$ was then modeled as

$$w[n] = g[n] \star b[n]$$

where $g[n]$ represents a mixed-phase, smooth, short-duration pulse and $b[n]$ represents a minimum-phase bubble oscillation with period T_{cb}, of the form

$$b[n] = \sum_{k=1}^{\infty} b[k]\delta[n - kT_b]$$

The trace's complex cepstrum $\hat{t}[n]$ is then of the form

$$\hat{t}[n] = \hat{g}[n] + \hat{b}[n] + \hat{m}[n] + \hat{r}[n]$$

where both $\hat{b}[n]$ and $\hat{m}[n]$ are periodic and causal, $\hat{g}[n]$ is concentrated around the origin, and $\hat{r}[n]$ has contributions only for times greater than T_1, the two-way travel time within the first earth layer. Buhl et al. showed that if there are N complex cepstral contributions of the water-column reverberation operator $m[n]$ in the interval $0 < n < T_1$, that is, if

$$NT_w \leq T_1 < (N + 1)T_w$$

where T_w denotes the reverberation period, then the elimination of these contributions by high-time cepstral gating will eliminate the first N reverberation multiples in the time domain and attenuate the remaining multiples to at most $1/(N + 1)$ of their original value. A similar argument holds for the elimination of the bubble-pulse oscillations. Thus, very significant dereverberation can be accomplished by this method in shallow-water situations with deep reflectors. Buhl et al. demonstrated the potential of this technique in the analysis of data recorded in 30 m of water. The results appear superior to those obtained by predictive deconvolution schemes.

A similar procedure was also investigated for deep-water dereverberation by Buttkus [43]. Here, of course, the cepstra of the reverberation trains will have long-period characteristics. He suggested, then, that dereverberation be made by notch gating the complex cepstrum at the multiples of T_w. The location of these multiples can often be directly estimated in the cepstral domain.

The use of homomorphic filters in seismic dereverberation was carefully evaluated by Gallemore [44] in both shallow- and deep-water situations, and their performance compared with the use of optimal tapped delay line filters, developed by Baggeroer [38], which may be equivalently represented in terms of predictive deconvolution filter structures. The criteria for evaluation were the percent of multiple energy removed, percent of reflector distortion, and visual improvement of the data when both methods were applied to synthetic noisy seismograms. The indications from this study were that homomorphic dereverberation appears to have greater potential in shallow-water applications, while predictive deconvolution appears to be more efficient for deep-water data.

Buttkus [43] has also discussed the limitations of seismic deconvolution by cepstral gating. Using experimental evidence, he concluded that the success of wavelet estimation by low-time gating was largely dependent on the signal-to-noise ratio of the input data, on the degree of overlapping of the cepstrum components of the source wavelet and of the reflectivity function, and on the choice of the low-time window.

Following Ulrych, Buttkus studied the influence of the noise on the unwrapped phase and attempted to prefilter the phase to minimize the influence of the noise. He also recommended that the seismic trace be linearly bandpass-filtered prior to homomorphic analysis. Finally, he concluded that the recovery of the reflector series by high-time gating the complex cepstrum of a seismogram should be discarded for cases where the noise is not negligibly small. He decided instead to proceed via homomorphic wavelet estimation to subsequently employ least-squares deconvolution by digital Wiener filtering.

For homomorphic wavelet estimation, he recognized that the use of a simple low-time gate on the cepstrum of the entire trace will usually have only limited success, since it often happens that the travel time within the first earth layer is short compared to the width of the seismic wavelet in the cepstral domain. Thus, the use of low-time gating may lead to large signal-estimation errors, because of the presence of the reflector series' low-time components.

Buttkus suggested instead a procedure for wavelet estimation based on the homomorphic analysis of a selected time interval of the seismic trace, with a length on the order of three to five times the wavelet length. The wavelet estimation was accomplished by comb gating the complex cepstrum. The determination of the wavelet arrival times and their positive linear combinations, necessary for the choice of the comb windows, was regarded as an echo-detection problem. Because of problems encountered with the influence of the noise in the phase curve, the estimation of the echo times was done by observation of spikes in the power cepstrum. This approach was tested under a variety of noise levels and in the presence of time-varying effects. Buttkus concluded that the main problem in the successful application of the homomorphic method was not the exact parameter determination of the arrival times or reverberation period but rather the implicit assumption that the data fit a time-invariant seismic model.

Finally, Otis and Smith [45] attempted to extend Stockham's blind deconvolution technique [46] to seismic processing. The method is also closely related to that of Clayton and Wiggins [40]. They considered a data base of several seismic traces, with a spatially stationary source pulse and spatially nonstationary reflector series. The complex cepstrum of each seismic record will be the sum of the cepstrum of the source pulse, which is nonvariable, plus the cepstrum of the reflector series, which varies from record to record. By averaging the complex cepstra of several records, an estimate of the nonvariable functions can be obtained, since the variable functions will tend to a mean value. Thus, the successful recovery of the seismic wavelet depends upon whether the mean of the reflector series cepstra is zero for all times. Based on experimental evidence, Otis and Smith suggested that this is the case whenever the geological structure changes at each shot point. Thus, by averaging a sufficiently large number of traces, an estimate of the complex cepstrum of the source pulse may be obtained.

This procedure has been referred to as log-spectral averaging or *cepstral stacking*. The estimated source wavelet is then used to construct a Wiener deconvolution filter.

We outlined above the basic strategies developed for full-band homomorphic analysis of time-invariant seismic models, which we refer to as time-invariant homomorphic analysis. A great number of seismic processing centers, both in the United States and abroad, have, in the past several years, indicated an interest in homomorphic analysis. Informal reports indicate that the method has been used with varying success.

In attempting to develop more reliable analysis strategies, we shall first investigate the limitations of time-invariant homomorphic analysis of time-varying, noisy seismic traces. This will be the topic of the next chapter.

VII.4 SUMMARY

Two classes of deconvolution filters were reviewed in this chapter: linear and homomorphic filters. Two types of linear filters were discussed, both based on parametric least-squares design methods. Wiener filters attempt to optimally convert the seismogram into the reflector series, and their design requires a priori estimation of the seismic wavelet. Predictive deconvolution filters are usually directed at seismic dereverberation. Their use as deconvolution filters is appropriate only when the seismic wavelet is minimum-phase.

A variety of homomorphic analysis methods with applications to teleseismic and seismic exploration were reviewed. All these techniques are based upon time-invariant seismic models. The seismic reflector series is assumed minimum-phase, this being ensured by exponentially weighting the trace, if necessary. Both wavelet estimation and seismic dereverberation by homomorphic filtering were discussed.

VIII

Time-invariant Homomorphic Analysis

VIII.1 INTRODUCTION

We reviewed in Chapter VII the various homomorphic analysis techniques that have been explored for seismic deconvolution. All these techniques are built upon a common analysis framework, which we shall refer to as *time-invariant homomorphic analysis.* This framework is essentially characterized by:

1. The representation of seismic traces in terms of time-invariant seismic models.
2. The use of exponential weighting, when necessary, to ensure the minimum-phase character of the seismic reflector series.

Within this framework, each technique is characterized by the specific method used to eliminate the undesired signal components in the complex cepstral domain: for example, low-time gating, comb gating, or cepstral stacking.

We investigate in this chapter the fundamental limitations that are associated inherently with time-invariant homomorphic analysis strategy. We demonstrate that this approach deterministically associates the low-time cepstral information with the data onset, thus constraining the wavelet estimates to be a replica of the first arrival. This result is most valuable in establishing the role of time-invariant homomorphic analysis in seismic data processing.

VIII.2 DETERMINISM AND TIME-INVARIANT HOMOMORPHIC ANALYSIS

As discussed in Chapter VI, seismic reflection traces may be represented as

$$s[n] = p[n] \star e[n] + \eta[n] \tag{8-1}$$

where $p[n]$ denotes the source pulse, $e[n]$ the impulse response of the earth, and $\eta[n]$ a noise component. It was also shown that $s[n]$ may be rewritten as

$$s[n] = \sum_{k=1}^{\infty} A_k w[n - n_k, \gamma_k] + \eta[n] \tag{8-2}$$

where $w[n, \gamma]$ represents a family of seismic wavelets. The corresponding seismic reflector series $r[n]$ was defined to be

$$r[n] = \sum_{k=1}^{\infty} A_k \delta[n - n_k] \tag{8-3}$$

Let

$$r'[n] = \beta^n r[n] \tag{8-4}$$

and

$$s'[n] = \beta^n s[n] \tag{8-5}$$

where $0 < \beta \leq 1$ is chosen so that $r'[n]$ is minimum-phase.

If we postulate the seismic trace to be modeled as the convolution of the reflector series with a seismic wavelet, then we are equivalently postulating the trace's complex cepstrum to be the sum of the cepstra of those two components.

In order to capture the fundamental limitations imposed by this analysis framework, let us assume that the first interarrival time, T_1, is arbitrarily large, so that the use of cepstral gating becomes most appropriate for the recovery of the signal components. For example, we may attempt to recover the reflector series using a high-time cepstral gate $g[n]$, defined as

$$g[n] = \begin{cases} 0, & n < T_1 \\ 1, & n \geq T_1 \end{cases} \tag{8-6}$$

Denoting by $r_g[n]$ the corresponding reflector series estimate such that

$$\hat{r}'_g[n] = \hat{s}'[n]g[n] \tag{8-7}$$

and noting that $r'_g[n]$ is minimum-phase, we may use the recursive relationship of equation (3-24) to show that

$$r_g[n] = r[n], \qquad n < n_2 \tag{8-8}$$

The corresponding wavelet estimate will be denoted as $w_g[n]$ and is such that

$$\hat{w}'_g[n] = \hat{s}'[n] - \hat{r}'_g[n] \tag{8-9}$$

so that

$$s[n] = w_g[n] \star r_g[n] \tag{8-10}$$

It follows, then, from equations (8-2), (8-3), and (8-8) that such wavelet estimate is constrained to *exactly match* the data onset; that is,

$$w_g[n] = s[n] = w[n - n_1, \gamma_1] + \eta[n], \qquad n < n_2 \tag{8-11}$$

This equation was derived using a specific gating procedure. It indicates, in general, however, that the redundant wavelet information contained in the later arrivals at $n = n_2, n_3, \ldots$ is *not* represented at all in the low-time values ($|n| < T_1$) of the complex cepstrum of the weighted seismic trace $s'[n]$. Thus, although the use of different time-invariant homomorphic wavelet estimation schemes (e.g., by low-time gating or cepstral stacking) may destroy the exact match as prescribed by equation (8-11), it is, nevertheless, a fact that such wavelet estimates will be determined only by the first wavelet arrival. This result is extremely important, in that it clearly identifies the inherently deterministic constraint imposed on any wavelet estimation scheme built upon the time-invariant homomorphic analysis framework. The practical consequences of this result depend on how closely the data fit the time-invariant model upon which the analysis is based. On data such that the seismic wavelet characteristics significantly change with time or on data severely corrupted by noise, it is to be expected that the determination of the first arrival will not be of much help in deconvolving the later arrivals.

VIII.3 EXAMPLES

We shall now illustrate the potential and limitations of time-invariant homomorphic analysis. Three types of synthetic seismograms will be analyzed. The first type, illustrated in Example 8.1, includes neither attenuation nor additive noise; the second type, illustrated in Example 8.2, includes attenuation but does not include additive noise. Finally, in Example 8.3, both attenuation and additive noise effects will be included.

In all these examples, the same earth structure was used, corresponding to a simple three-layer model with a two-way travel time of 300 ms for the top layer and 500 ms for the middle layer. This earth is illuminated from above by a seismic source, as in marine profiling. Water-layer reverberation effects are not included. The sampling interval was set at 2.44 ms. The corresponding reflector series is illustrated in Figure 8.1c.

In this analysis, symmetrical low-time and high-time cepstral gating will be used instead of the specific gating procedures described in the preceding section, to illustrate the kind of deviations to be expected from the exact matching of the wavelet estimate to the data onset, as discussed above.

EXAMPLE 8.1

In this example we shall analyze the synthetic seismogram $s[n]$ of Figure 8.1a, which corresponds to the convolution of the airgun signature $p[n]$ of Figure 8.1b with the earth impulse response $e[n]$ of Figure 8.1c. This impulse response corresponds to the earth model described above, assuming no loss mechanism; that is, the layer attenuation coefficients were set to zero. As a consequence, the earth impulse response $e[n]$ equals the reflector series $r[n]$ associated with this earth model.

The seismogram $s[n]$ was exponentially weighted with a weighting equal to 0.99 to yield a weighted seismogram $s'[n]$. Figure 8.1d depicts the complex cepstrum $\hat{s}'[n]$ of the weighted seismogram, and Figures 8.1e and f depict the low-time and high-time components of $\hat{s}'[n]$, respectively, denoted by $\hat{s}'^L[n]$ and $\hat{s}'^H[n]$. The cepstral cutoff time was set at 293 ms. Figures 8.1g and h depict the time sequences associated with such low-time and high-time components, $s'^L[n]$ and $s'^H[n]$. Finally, after exponential deweighting, the results of homomorphic analysis of the seismic trace of Figure 8.1a, $s^L[n]$ $s^H[n]$, are shown in Figures 8.1j and l. For comparison purposes the airgun pulse $p[n]$ and reflector series $r[n]$ are depicted in Figures 8.1i and k.

EXAMPLE 8.2

In this example we shall analyze the synthetic seismogram $s[n]$ of Figure 8.2a, which corresponds to the convolution of the airgun signature $p[n]$ of Figure 8.2b, with the earth impulse response $e[n]$ of Figure 8.2c. This time the attenuation coefficients were set to 0.001 for the first layer and 0.002 for the second layer. The analysis proceeds as in Example 8.1, using exactly the same weighting factor and cepstral cutoff time. The results are depicted in Figures 8.2d through 8.2l.

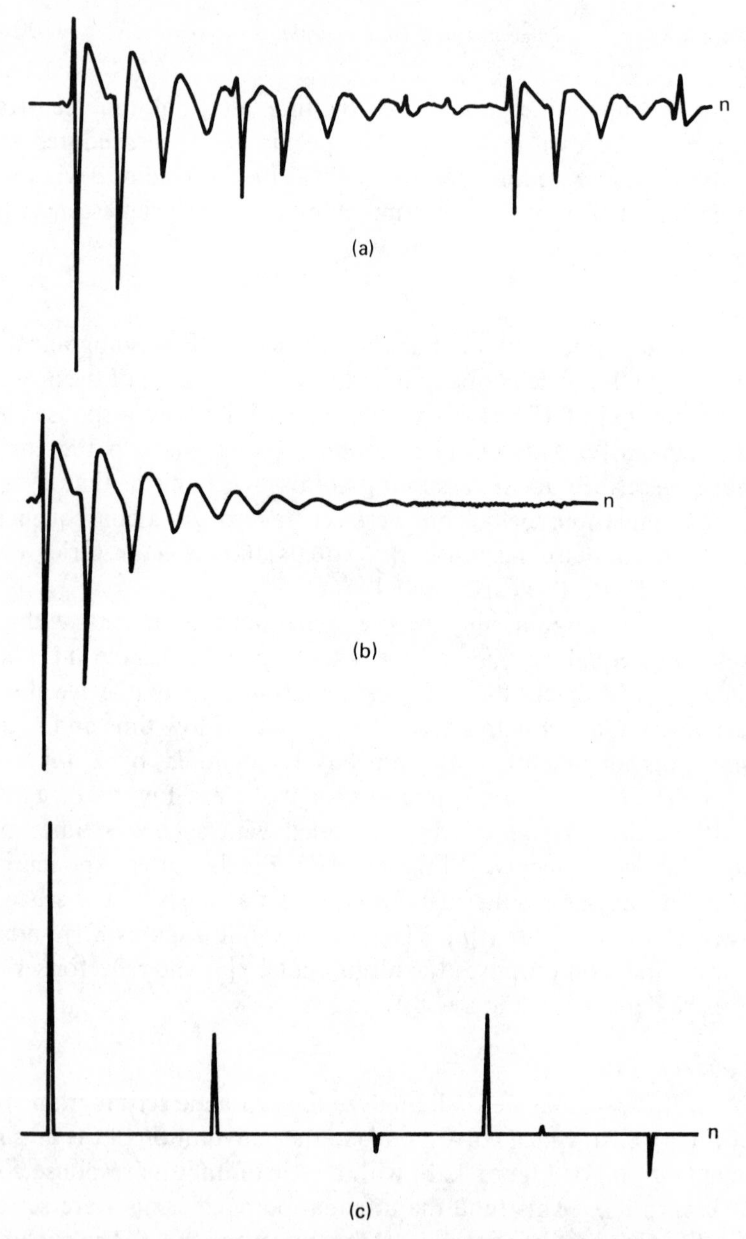

Figure 8.1 Time-invariant homomorphic analysis of synthetic seismogram (noise and attenuation effects not included): (a) synthetic trace $s[n]$; (b) source pulse $p[n]$; (c) Earth impulse response $e[n]$

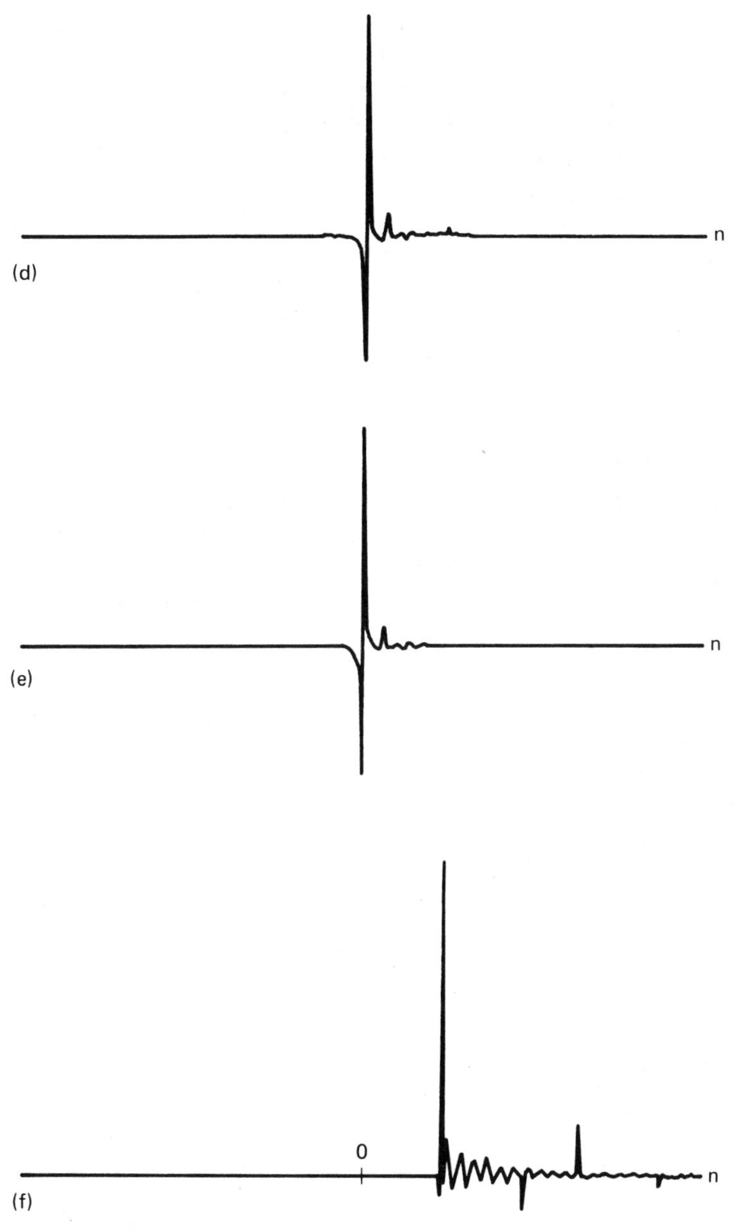

Figure 8.1 (Cont.): (d) complex cepstrum of exponentially weighted trace $\hat{s}'[n]$; (e) low-time cepstral component $\hat{s}'^{L}[n]$; (f) high-time cepstral component $\hat{s}'^{H}[n]$ (the vertical scale has been magnified × 57)

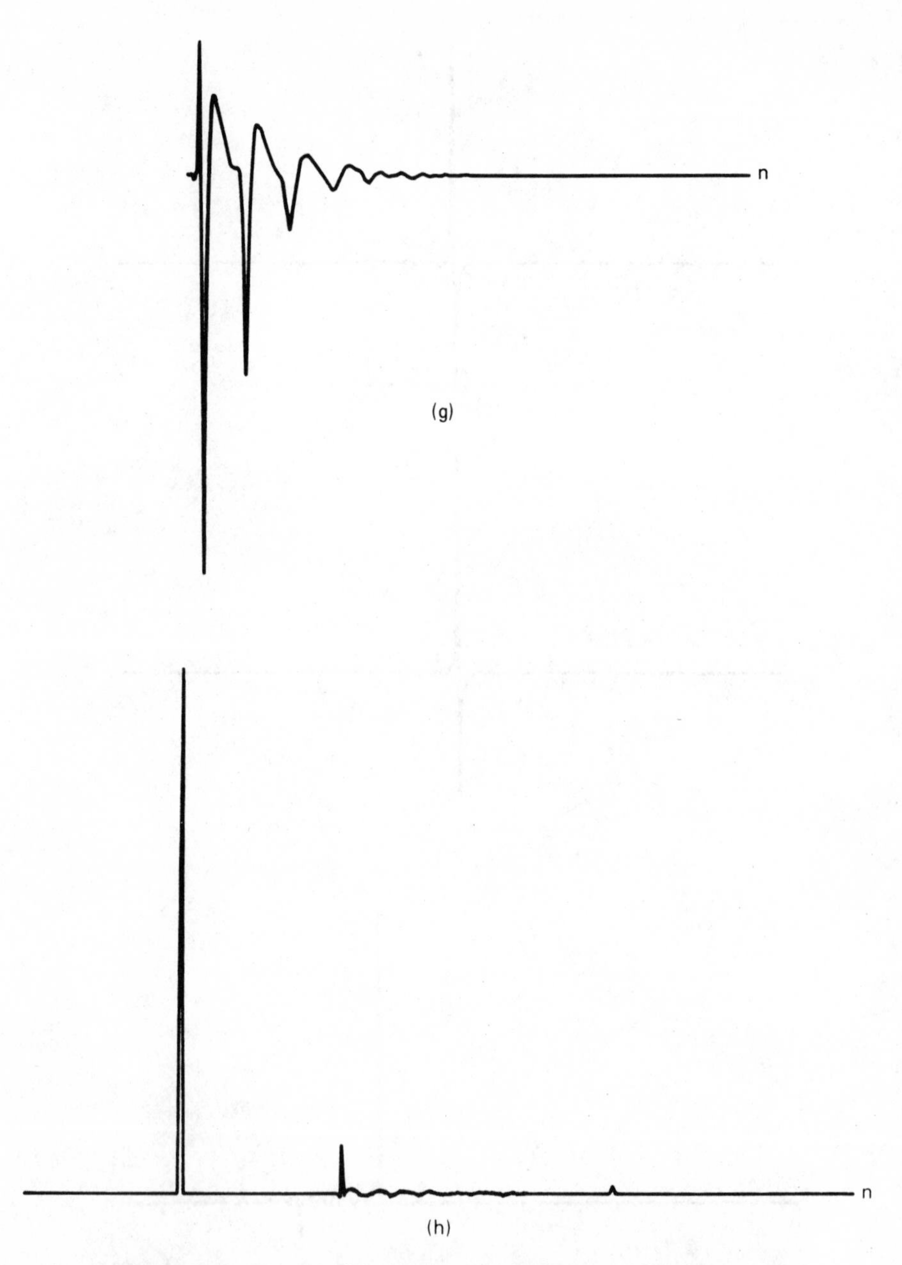

Figure 8.1 (Cont.): (g) low-time component of weighted trace $s'^L[n]$; (h) high-time component of weighted trace $s'^H[n]$

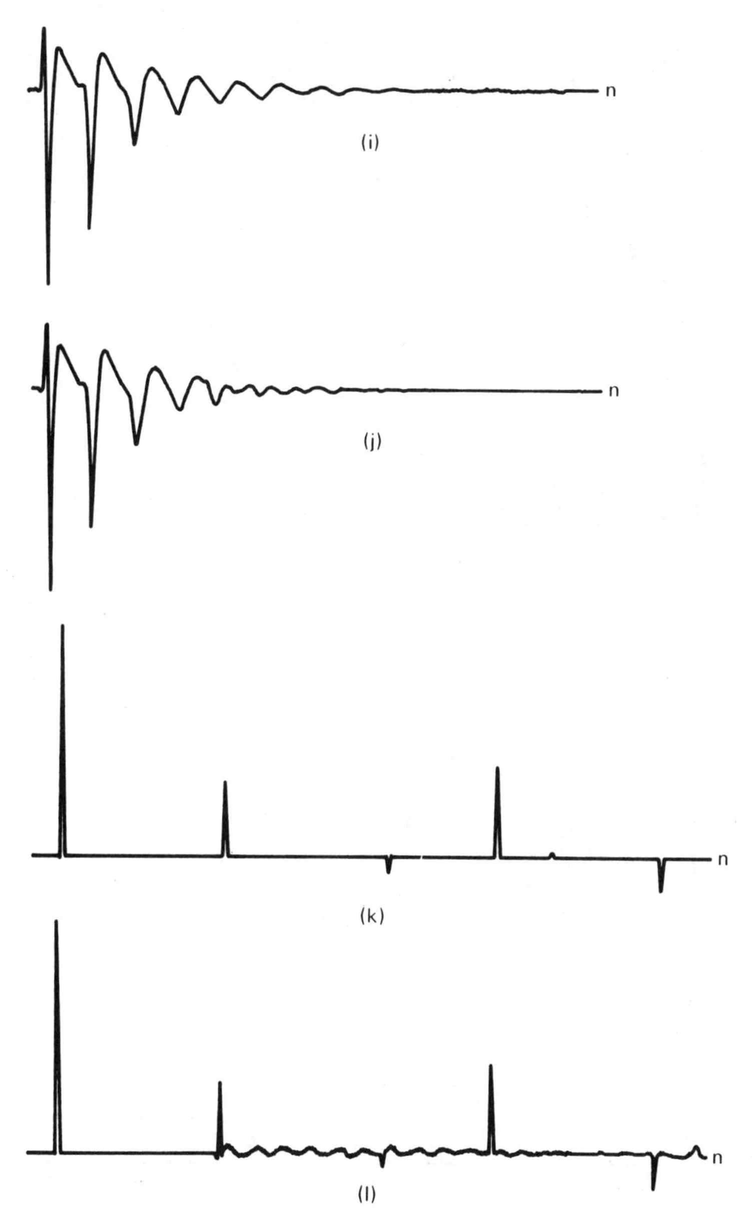

Figure 8.1 (Cont.) : (i) source pulse $p[n]$; (j) deweighted low-time component $s^L[n]$; (k) reflector series $r[n]$; (l) deweighted high-time component $s^H[n]$

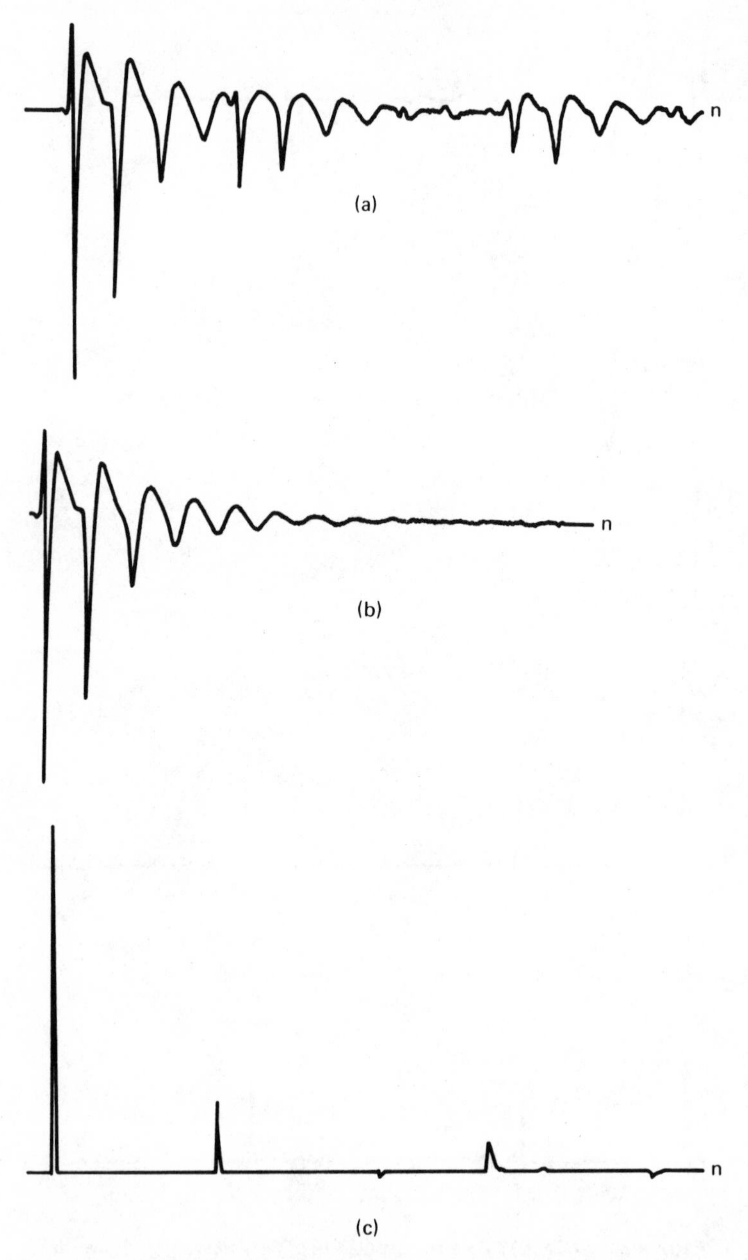

Figure 8.2 Time-invariant homomorphic analysis of synthetic seismogram (includes attenuation effects) : (a) synthetic trace $s[n]$; (b) source pulse $p[n]$; (c) Earth impulse response $e[n]$

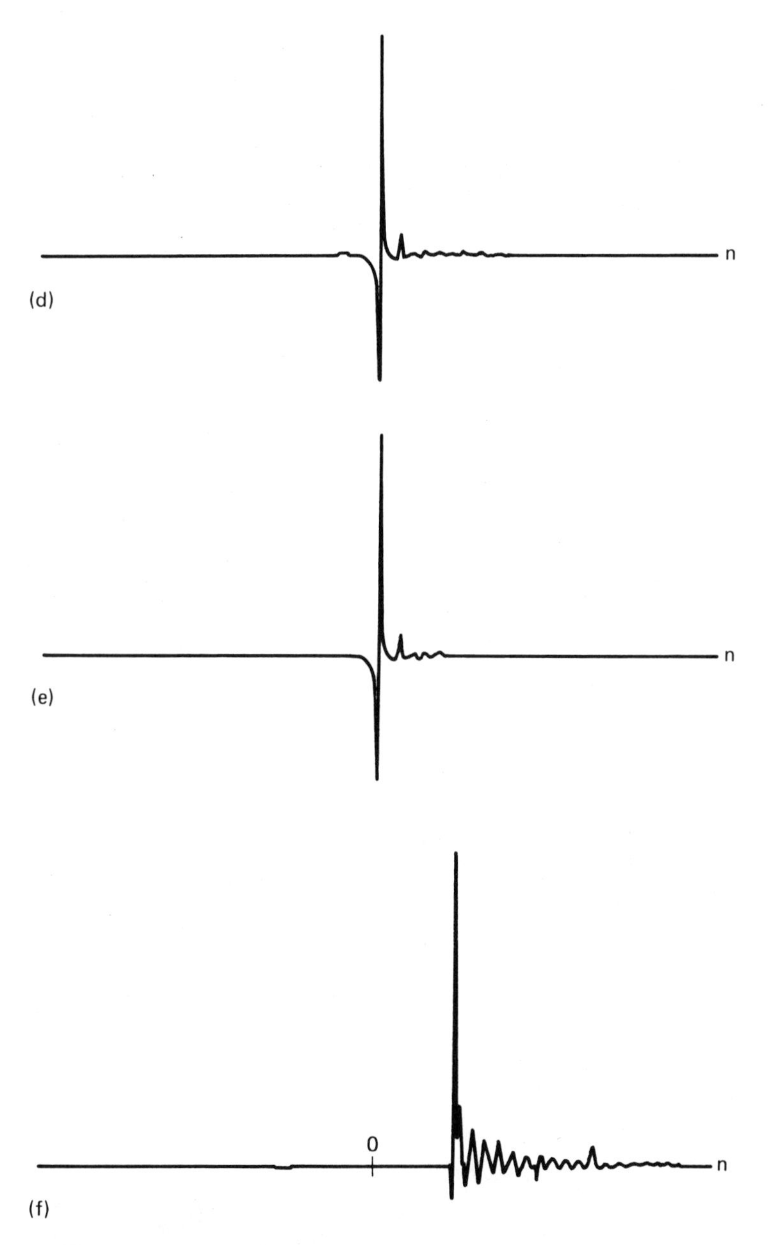

Figure 8.2 (Cont.) : (d) complex cepstrum of exponentially weighted trace $\hat{s}'[n]$; (e) low-time cepstral component $\hat{s}'^{L}[n]$; (f) high-time cepstral component $\hat{s}'^{H}[n]$ (the vertical scale has been magnified × 57)

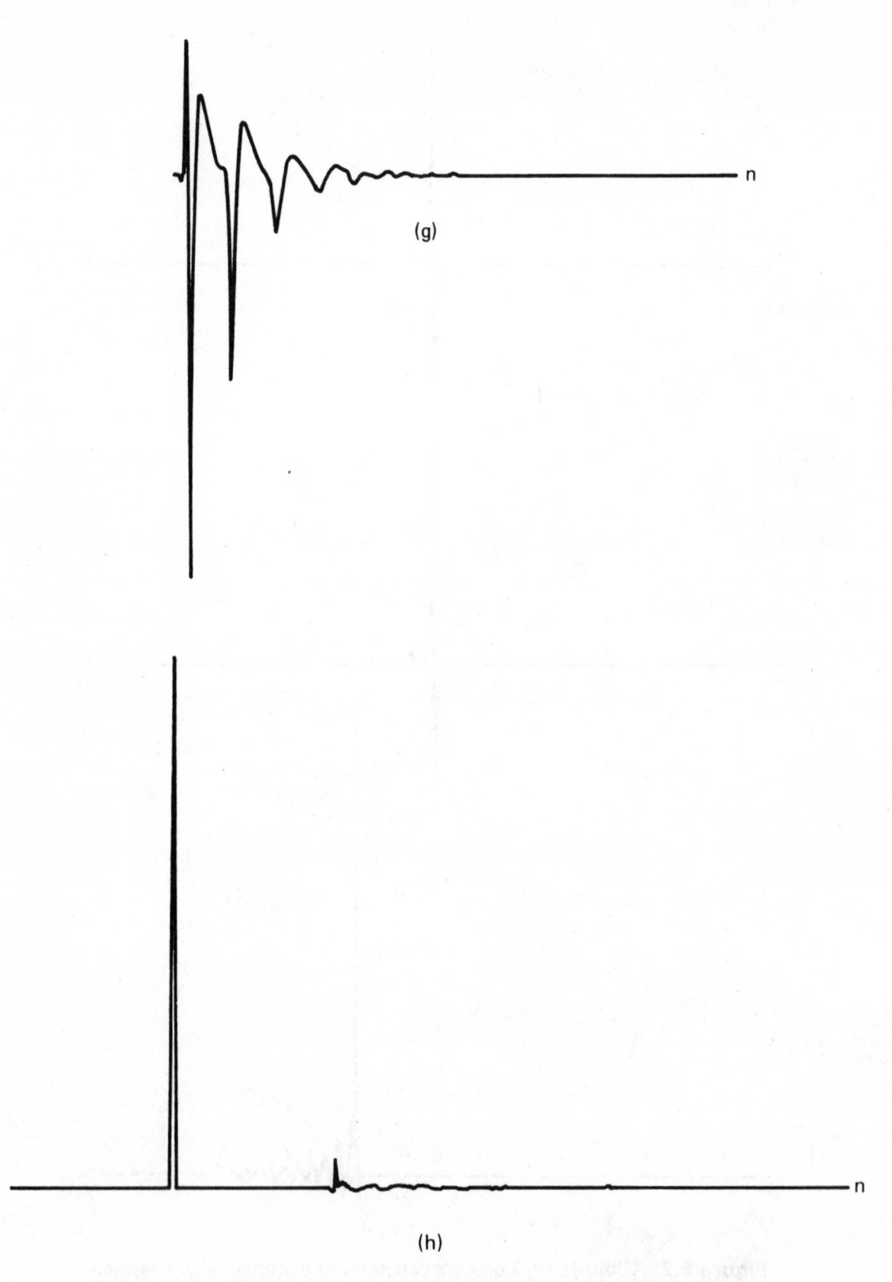

Figure 8.2 (Cont.): (g) low-time component of weighted trace $s'^L[n]$; (h) high-time component of weighted trace $s'^H[n]$

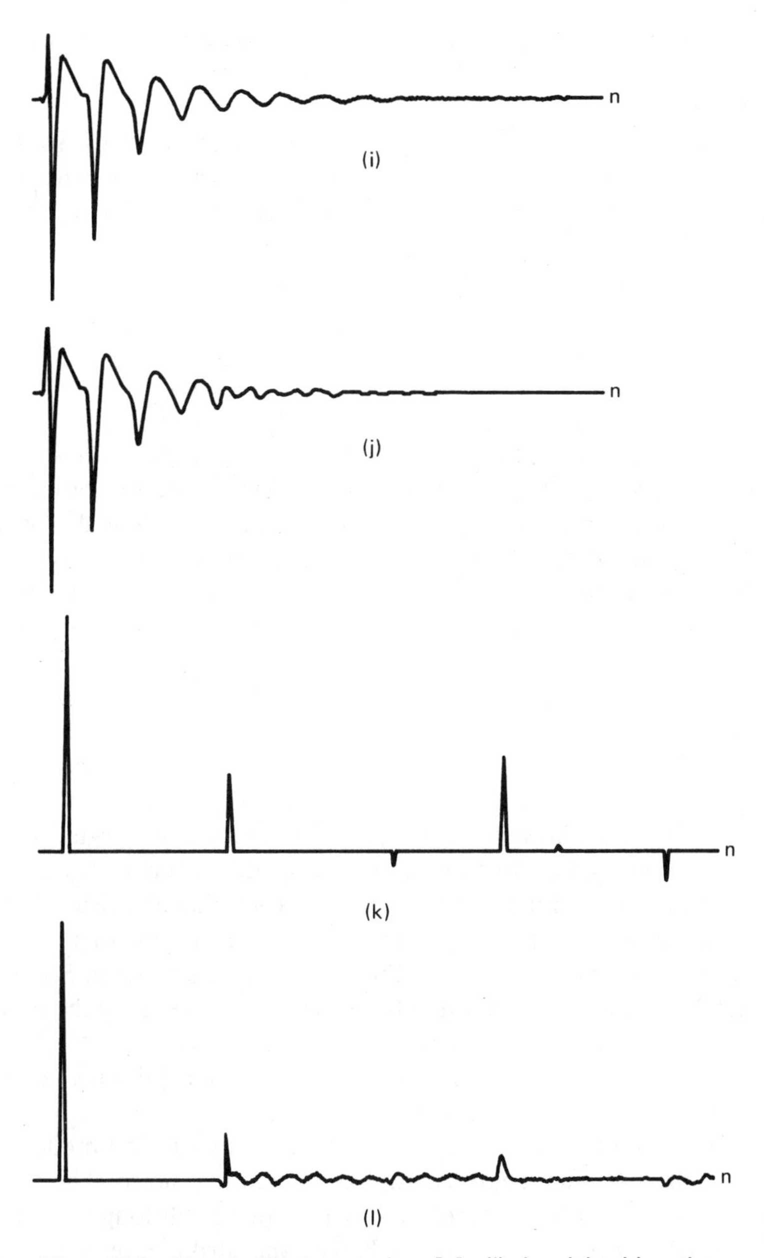

Figure 8.2 (Cont.) : (i) source pulse $p[n]$; (j) deweighted low-time component $s^L[n]$; (k) reflector series $r[n]$; (l) deweighted high-time component $s^H[n]$

EXAMPLE 8.3

Finally, let us consider the seismogram of Example 8.2, shown in Figure 8.3c, corrupted by an additive noise component $\eta[n]$ depicted in Figure 8.3b. The SNR was set at +10 dB. The resulting noisy seismogram is shown in Figure 8.3a. Following the same analysis procedure as in the previous examples, the noisy seismic trace was analyzed, as illustrated in Figures 8.3d through l.

The examples shown above exhibit a number of important characteristics. Example 8.1 corresponds to a truly time-invariant seismic model. The most striking feature observed in this example was the ability to recover the seismic pulse by means of low-time cepstral gating. The recovery of the seismic reflector series by means of high-time gating is also of high-quality. Note the presence of a noise component in both estimates, which is observed after the cutoff time T_c and is due to the presence of a nonzero high-time cepstral component associated with the seismic wavelet, which is overlapping that of the reflector series.

In Example 8.2 we are no longer in the presence of a time-invariant model, since now the earth reflector series is distinct from the earth impulse response. Thus, the characteristics of the seismic wavelet change in time as a result of the attenuation within the earth. Note, however, the recovery of the seismic airgun pulse by low-time gating. In fact, observe that this estimate is exactly identical to the estimate of Example 8.1, even though arising from the analysis of different seismograms. This result was to be expected, however, in light of the considerations made in Section VIII.2, since the first arrival in both examples has exactly the same shape, which is precisely that of the airgun pulse.

The high-time estimate is now seen to match $e[n]$, thus failing to estimate the reflector series $r[n]$.

In Example 8.3 we are analyzing the time-varying model of Example 8.2, after being corrupted by additive noise. The low-time estimate still captures the airgun pulse, although it now appears corrupted by noise. Observe the striking resemblance between the low-time estimate and the first arrival in Figure 8.3a. This resemblance becomes practically an identity after exponential deweighting, as depicted in Figure 8.3j.

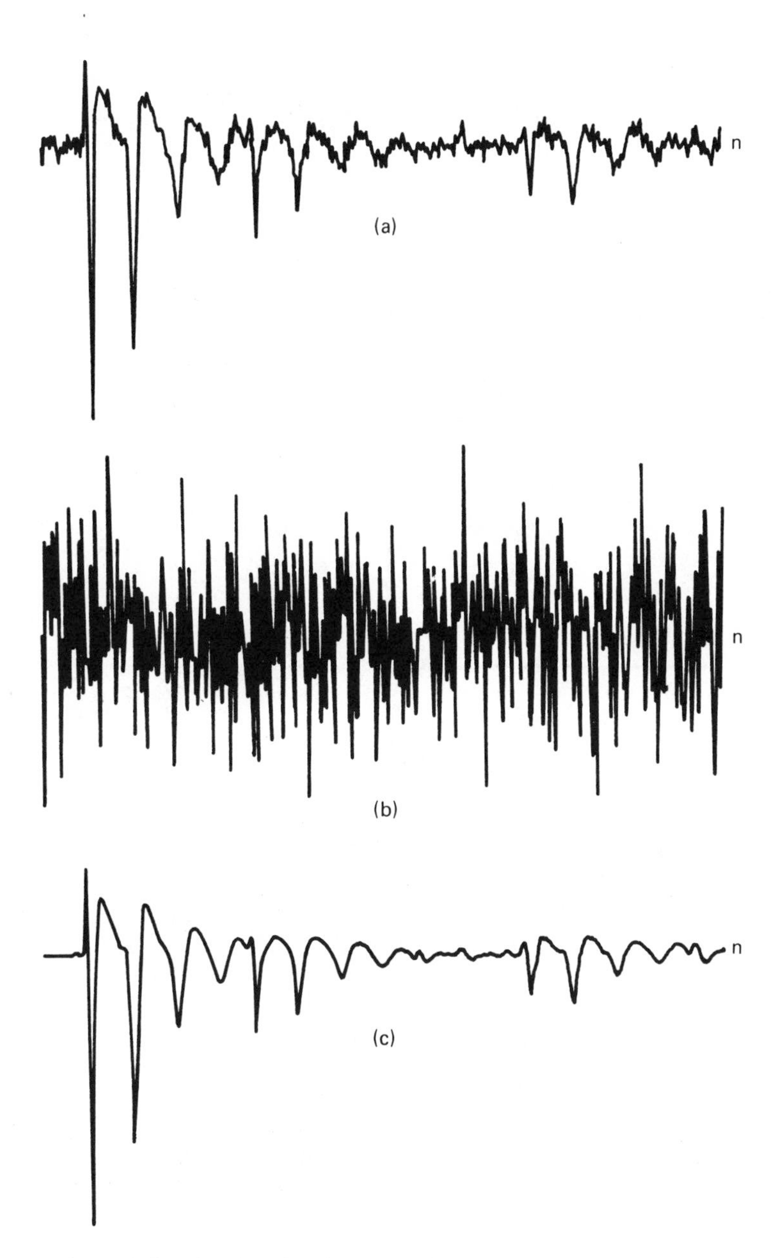

Figure 8.3 Time-invariant homomorphic analysis of synthetic seismogram (includes both noise and attenuation effects): (a) noisy synthetic trace $s[n]$; (b) noise component $\eta[n]$ (the vertical scale has been magnified $\times$ 10); (c) noiseless synthetic trace $p[n] \star e[n]$

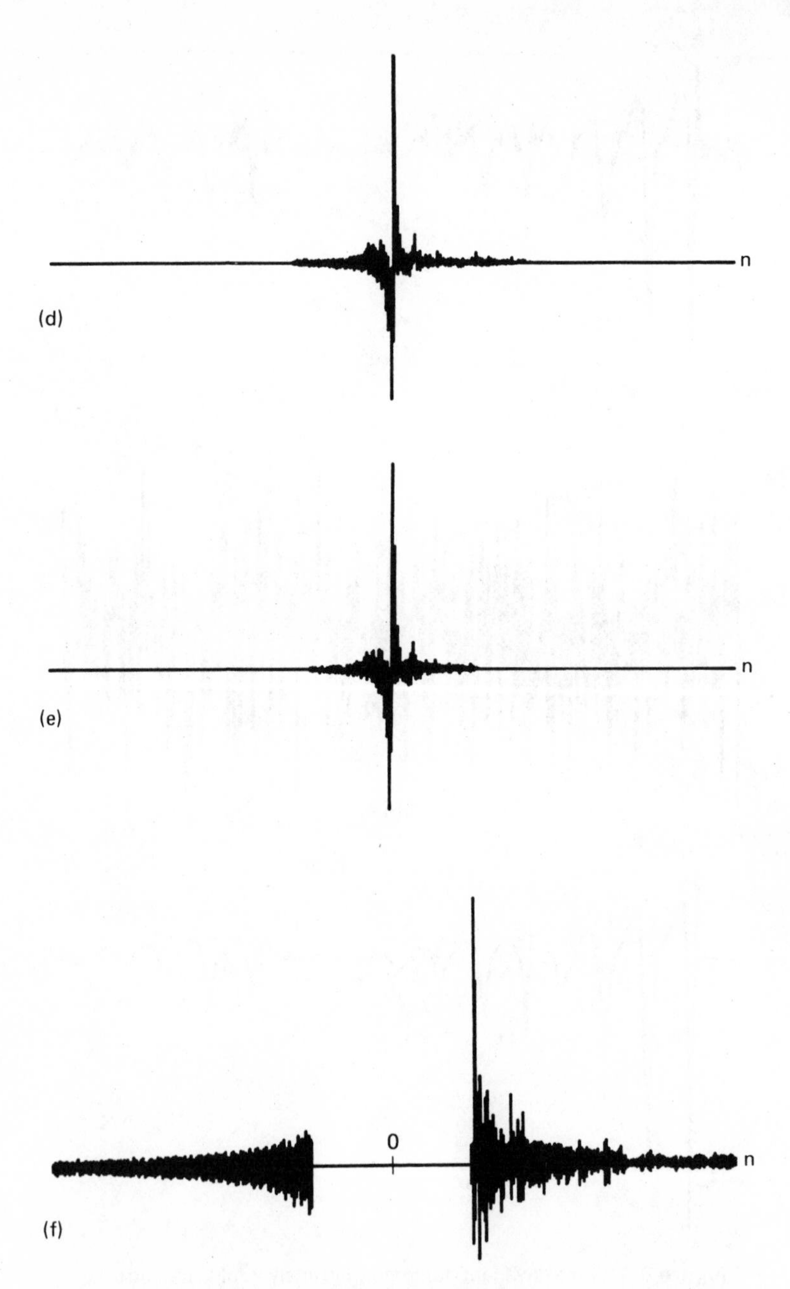

Figure 8.3 (Cont.): (d) complex cepstrum of exponentially weighted trace $\hat{s}'[n]$; (e) low-time cepstral component $\hat{s}'^{L}[n]$; (f) high-time cepstral component $\hat{s}'^{H}[n]$ (the vertical scale has been magnified × 57)

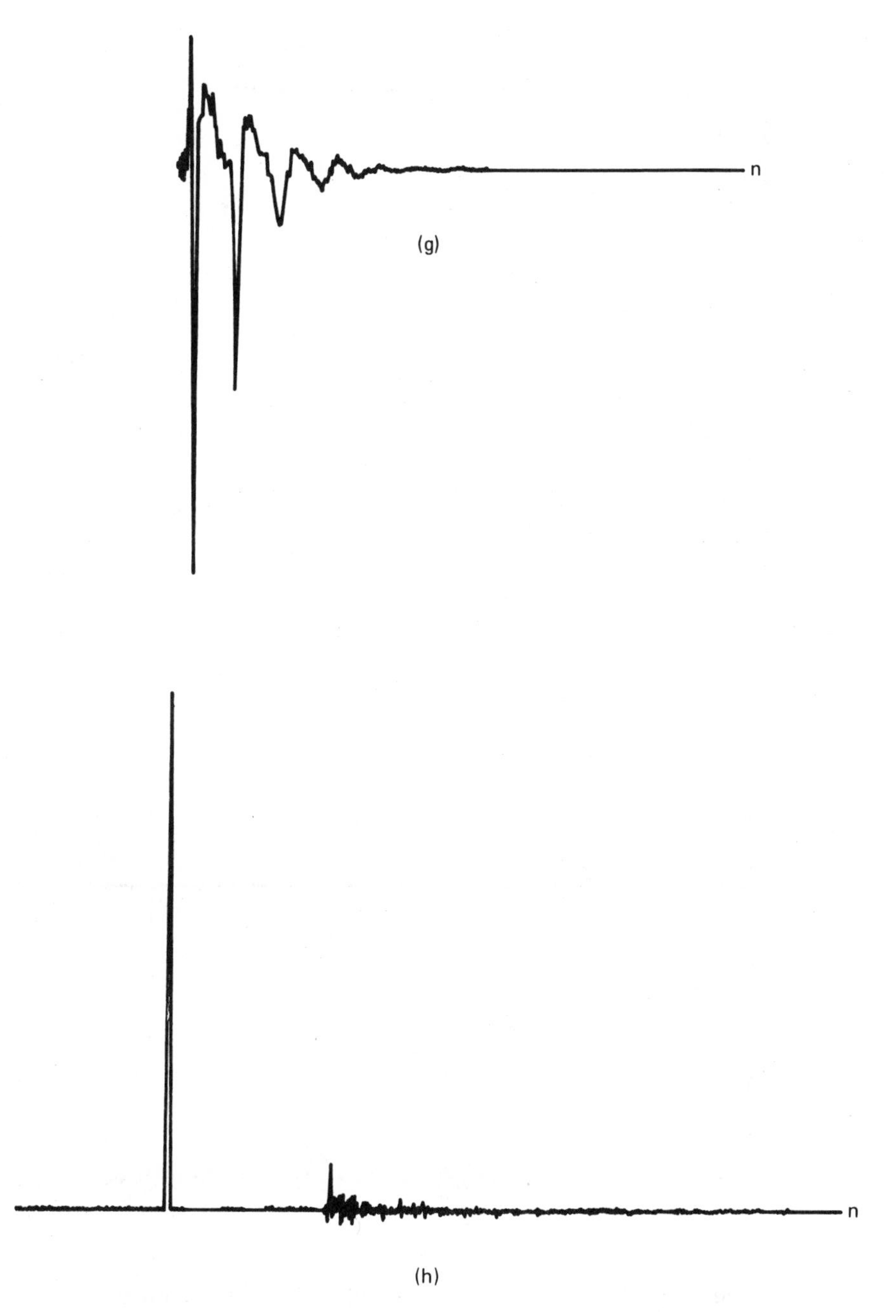

Figure 8.3 (Cont.): (g) low-time component of weighted trace $s'^L[n]$; (h) high-time component of weighted trace $s'^H[n]$

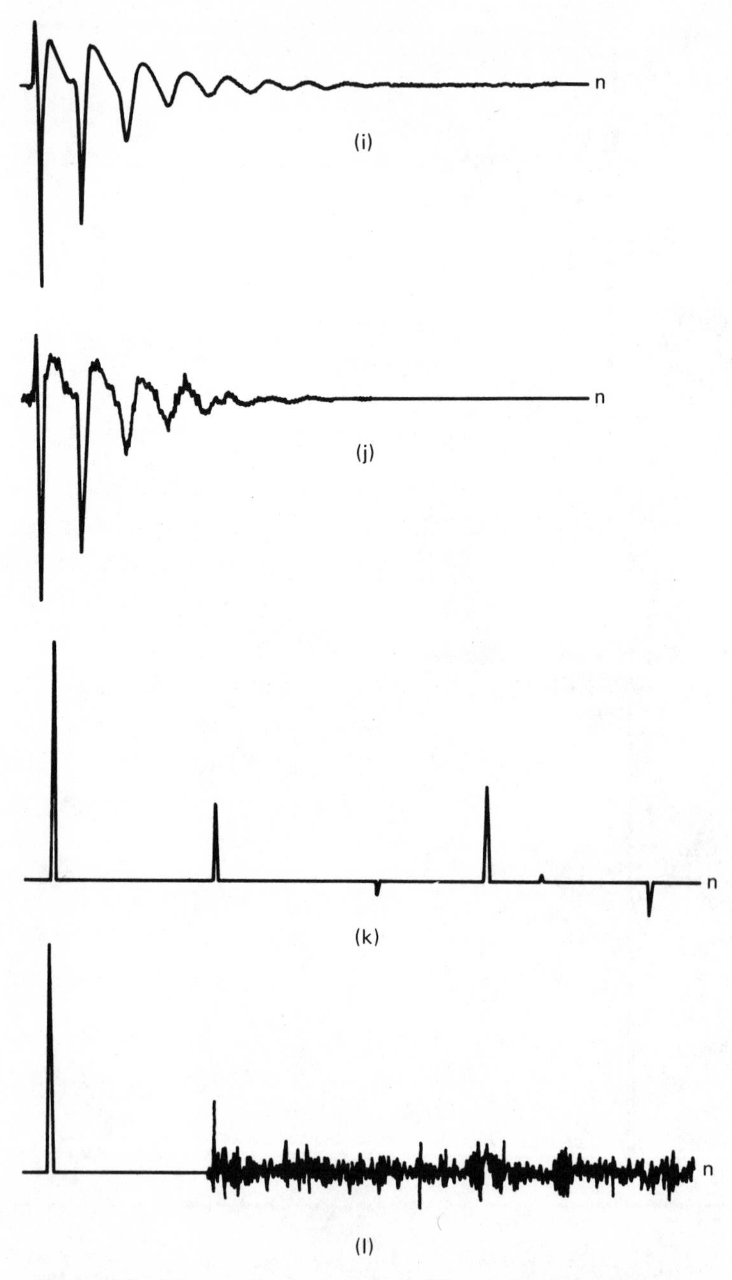

Figure 8.3 (Cont.): (i) source pulse $p[n]$; (*j*) deweighted low-time component $s^L[n]$; (k) reflector series $r[n]$; (l) deweighted high-time component $s^H[n]$

The estimated reflector series does indeed confirm such identity, since the first reflector has been perfectly resolved, as seen in Figure 8.3l, in good agreement with the results of Section VIII.2, that is, equation (8-8). However, the reflector information after the cutoff time T_c is almost hopelessly immersed in noise, thus rendering such an estimate basically useless.

VIII.4 ROLE OF TIME-INVARIANT HOMOMORPHIC ANALYSIS IN SEISMIC DATA PROCESSING

We showed in Section VIII.2 that, within a time-invariant homomorphic analysis fremework, only the first wavelet arrival is represented in the low-time values of a seismic trace's complex cepstrum. Thus, the high-time values are not only associated with the reflector series but also with any deviation between the later arrivals and the first that is due either to attenuation or noise. In general, then, the reflector series estimate by high-time gating has a perfectly resolved first arrival, the remainder being corrupted by noise, with a severity that depends on the particular time-varying nature of the data and on the signal-to-noise ratio. In many cases, the estimates are so severely corrupted that they are of little value.

Note, however, that by time-invariant homomorphic analysis, using either low-time cepstral gating or cepstral stacking as required, an estimate of the *first* arrival may be obtained, without a priori requirements on the wavelet shape or without requiring the reflector series to be uncorrelated.

It is thus within this framework that the role of time-invariant homomorphic analysis in seismic processing must be established. In particular, it may be used in teleseismic analysis, since the goal is primarily the determination of the wavelet shape, which is basically stationary in time. This approach may also be appropriate within a seismic dereverberation context, by exploiting the periodicity of the reverberation operator, as already discussed in Chapter VII.

The use of time-invariant homomorphic analysis in the deconvolution of reflection seismograms is, however, in general not recommended, since the determination of the shape of the first arrival will, in most cases, be of little help in resolving the deeper reflectors. Thus, the estimation of the seismic reflector series using homomorphic signal processing requires in general a different analysis strategy, which takes into account the specific time-varying characteristics of the trace. The development of a framework for such analysis and the investigation of appropriate homomorphic filtering strategies will be the topic of the next two chapters.

VIII.5 SUMMARY

Time-invariant homomorphic analysis techniques are characterized by the fact that the data, which are represented in terms of time-invariant seismic models, are exponentially weighted as necessary, in order to ensure the minimum-phase character of the reflector series.

Time-invariant homomorphic analysis techniques are fundamentally limited by the fact that the low-time values of the trace's complex cepstrum are uniquely determined by the first wavelet arrival and depend in no way on subsequent arrivals.

This fact is a major limitation to the use of any technique built upon such time-invariant homomorphic analysis framework in the analysis of seismic reflection data.

IX

Short-time Homomorphic Analysis: Basic Framework

IX.1 INTRODUCTION

The preceding chapter was dedicated to the study of a class of homomorphic seismic processors that attempt to remove the wavelet effects from a seismic trace, assuming stationarity of the pulse shape in time.

As discussed in Chapter VI, such stationarity does not hold in general. However, the change in wavelet character is essentially a slowly time-varying phenomenon. Motivated by this fact, we shall develop an analysis framework based on a segmentation of the seismic trace into short sections, each being represented in terms of a *windowed* time-invariant seismic model. We shall then investigate the principal considerations in the homomorphic analysis of such short-time seismic segments. In particular, we study the short-time windowing effects, both in time and in the cepstral domains. Specific short-time analysis strategies will be discussed in Chapter X.

IX.2 SHORT-TIME MODELS

In seeking a simplified model representation for the seismic trace, we shall start by analyzing the attenuation effects of the earth in regard to the variation in time of the wavelet characteristics. As the direct observation of seismic data shows, there may be a noticeable difference between the broad-band wavelets arising from the shallow reflectors and the narrow-band wavelets arising from the deeper reflectors. This is simply a confirmation of the progressive low-pass filtering effect of the earth as the seismic wave travels farther and farther into the earth, as predicted in equation (6-3).

The change in wavelet character does come about, however gradually in time, so that, on a short-time basis, all seismic wavelets display very similar shapes. This motivates short-time analysis schemes, where the data are segmented into short sections, each one being modeled in a time-invariant manner. Thus, the overall time-varying behavior is modeled as a sequence of time-invariant models, each being valid within a particular short-time interval.

Let us now consider in detail a framework for short-time analysis. We denote:

1. The kth short-time interval Δ_k as

$$\Delta_k = \left(N_k - \frac{L_k}{2}, N_k + \frac{L_k}{2}\right) \tag{9-1}$$

   where N_k is the center of the interval and L_k its length.
2. The kth data segment $s[n; \Delta_k]$ as

$$s[n; \Delta_k] = \begin{cases} s[n], & n \in \Delta_k \\ 0, & \text{elsewhere} \end{cases} \tag{9-2}$$

3. The kth reflector series segment $r[n; \Delta_k]$ as

$$r[n; \Delta_k] = \begin{cases} r[n], & n \in \Delta_k \\ 0, & \text{elsewhere} \end{cases} \tag{9-3}$$

4. The kth seismic wavelet $w[n; \Delta_k]$ as

$$w[n; \Delta_k] = w[n, \gamma_k] \tag{9-4}$$

   where γ_k represents the lumped attenuation coefficient associated with the arrivals within Δ_k.

If we denote by $g[n; \Delta_k]$ a rectangular window such that

$$g[n; \Delta_k] = \begin{cases} 1, & n \in \Delta_k \\ 0, & \text{elsewhere} \end{cases} \tag{9-5}$$

we may then precisely formulate the fact that the wavelet character does not significantly change within Δ_k by writing

$$s[n; \Delta_k] = [w[n; \Delta_k] \star r[n; \Delta_k]]g[n; \Delta_k] \tag{9-6}$$

That is, the kth data segment essentially satisfies a time-invariant model, except that only a short-time data interval Δ_k is available.

More generally, we may segment the data using a variety of short-time windows. Letting $g_i[n; \Delta_k]$ denote a short-time window defined in Δ_k, where the subscript i will be used to refer to the particular window used, we shall represent the corresponding windowed seismic segment as $s_i[n; \Delta_k]$, that is,

$$s_i[n; \Delta_k] = s[n; \Delta_k]g_i[n; \Delta_k] \tag{9-7}$$

This short-time framework is illustrated in Figure 9.1a. Within this framework, then, the basic analysis goal is the recovery of the seismic reflector segment $r[n; \Delta_k]$, $\forall\, k$. The overall seismic processor is thus constituted by a series of analysis operations, each being activated on a particular time interval, as illustrated in Figure 9.1b. Our purpose is now to investigate the role of homomorphic filtering in the analysis of short-time seismic segments.

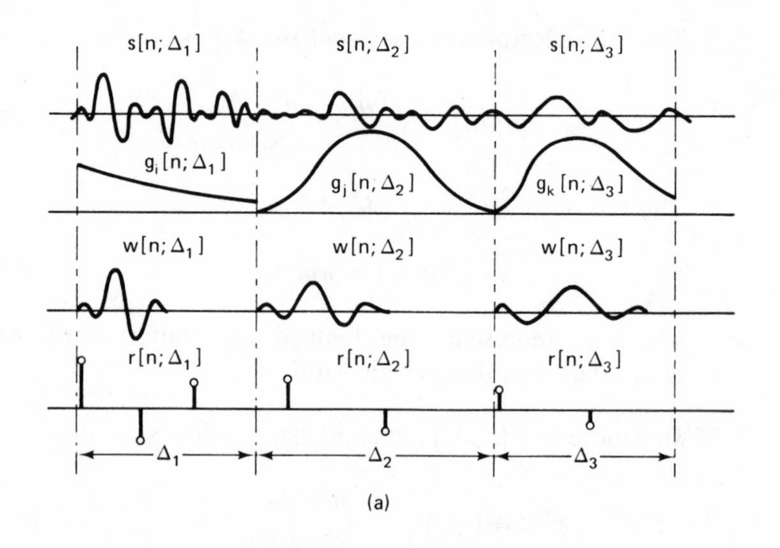

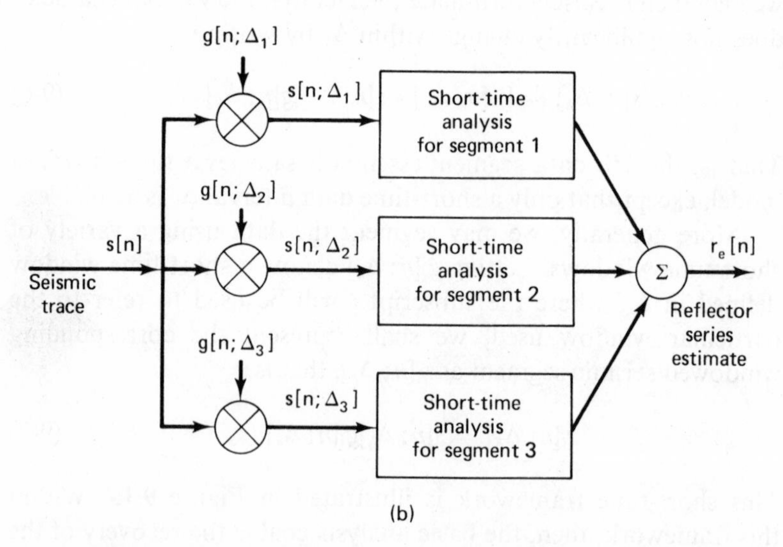

Figure 9.1 Short-time analysis: (a) signal models; (b) signal processor

IX.3 BANDPASS MAPPING

The homomorphic analysis of a seismic segment must take into account the bandpass nature of the data. In particular, the passband must be shifted and stretched to occupy the entire frequency band, by prefiltering the data through a bandpass-mapping system BP.

The major effects of this operation on signals that can be represented as a convolution of a wavelet with an impulse train have already been discussed and illustrated in Chapters IV and V. It was shown [see equation (5-25)] that the bandpass-mapped data can also be represented as a convolution of a wavelet with an impulse train, whose arrival times and amplitudes are simply related to the input wavelet arrival times and amplitudes. It was noted, however, that the polarity of these arrivals was not preserved through the transformation. We may thus model a bandpass-mapped seismic segment in a manner similar to that of Section IX.2. In particular, denoting by $\tilde{\Delta}_k$ the time interval occupied by a seismic segment $s[n; \Delta_k]$ after bandpass mapping, and representing by $\tilde{s}[n; \tilde{\Delta}_k]$ the corresponding bandpass-mapped segment, we have

$$\tilde{s}[n; \tilde{\Delta}_k] = [\tilde{w}[n; \tilde{\Delta}_k] \star \tilde{r}[n; \tilde{\Delta}_k]]\tilde{g}[n; \tilde{\Delta}_k] \qquad (9\text{-}8)$$

where $\tilde{w}[n; \tilde{\Delta}_k]$ denotes the basic wavelet pulse within $\tilde{\Delta}_k$ and $\tilde{r}[n; \tilde{\Delta}_k]$ is an impulse train that represents the various wavelet arrival times and amplitudes. The sequence $\tilde{g}[n; \tilde{\Delta}_k]$ is a rectangular window on $\tilde{\Delta}_k$. The wavelet $\tilde{w}[n; \tilde{\Delta}_k]$, as defined above, is related to the bandpass-mapped input wavelet $w[n; \Delta_k]$ as follows:

$$\tilde{w}[n; \tilde{\Delta}_k] \simeq \mathrm{BP}\{w[n; \Delta_k]\} \star \tilde{b}[n] \qquad (9\text{-}9)$$

where $\tilde{b}[n]$ represents the envelope of the responses of BP to impulses $\delta[n - k]$. The impulse train $\tilde{r}[n; \tilde{\Delta}_k]$ is related to the reflector segment as

$$\mathrm{BP}\{r[n; \Delta_k]\} \simeq \tilde{r}[n; \tilde{\Delta}_k] \star \tilde{b}[n] \qquad (9\text{-}10)$$

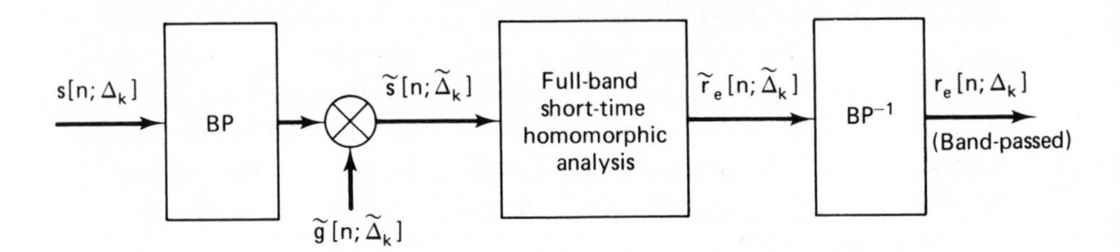

Figure 9.2 Short-time bandpass homomorphic analysis

Within this context, thus, the purpose of the homomorphic analysis of $\tilde{s}[n; \tilde{\Delta}_k]$ is the recovery of $\tilde{r}[n; \tilde{\Delta}_k]$, which must be subsequently filtered through $(\text{BP})^{-1}$ to yield a bandpass estimate of the reflector segment $r[n; \Delta_k]$, as illustrated in Figure 9.2. If necessary, the estimation of the out-of-band components of $r[n; \Delta_k]$ from such bandpass estimate may be accomplished by means of parametric restoration techniques [47].

Keeping in mind the need for the bandpass-mapping operation prior to full-band homomorphic analysis, we shall henceforth take advantage of the similarity of the models of equations (9-6) and (9-8) and proceed with our investigation by assuming the seismic segment to have full-band characteristics.

IX.4 SHORT-TIME WINDOWING EFFECTS

In homomorphic signal analysis two aspects must be carefully considered:

1. The representation of the signal to be analyzed in terms of a convolution of components, one of which constitutes or is simply associated with the information we wish to recover.
2. The specific properties of the signal components in the cepstral domain.

Our first task shall then be to develop a convolutional representation for a seismic segment $s[n; \Delta_k]$. At issue is, of course, the distortion introduced in the underlying convolutional model of equation (9-6), $w[n; \Delta_k] \star r[n; \Delta_k]$, by the truncation of the signal at the extremes of the interval Δ_k. A similar problem occurs in short-time spectral analysis, where the approach has been to multiply the data segment by a window $g_i[n; \Delta_k]$, such as a Hamming window, having more gradual onset and offset transitions in order to minimize the transition effects at the beginning and end of the short-time interval. Within this context it is often assumed that if the window varies slowly with respect to the seismic wavelet, the distortion introduced by the window can be basically accounted for in terms of a change in the amplitudes of the wavelet arrivals as follows:

$$s_i[n; \Delta_k] \simeq w[n; \Delta_k] \star r_i[n; \Delta_k] \tag{9-11}$$

where the component $r_i[n; \Delta_k]$ represents the windowed reflector segment:

$$r_i[n; \Delta_k] = r[n; \Delta_k]g_i[n; \Delta_k] \tag{9-12}$$

In terms of such a model, then, the complex cepstrum of the windowed segment $s_i[n; \Delta_k]$ can be represented as the sum of two components, one, $\hat{w}[n; \Delta_k]$, being independent of the particular window chosen and of its positioning in time within Δ_k, the other, $\hat{r}_i[n; \Delta_k]$, being determined by the windowed reflector segment.

In a first attempt to investigate short-time windowing effects in the context of homomorphic analysis [48], some pertinent questions were raised concerning the adoption of the model of equation (9-11), to generally describe the complex cepstrum of a short-time segment. We shall investigate this subject further, by adopting a rigorous representation of a short-time windowed segment as

$$s_i[n; \Delta_k] = w_i[n; \Delta_k] \star r_i[n; \Delta_k] \tag{9-13}$$

where $w_i[n; \Delta_k]$ is now defined in the frequency domain as

$$W_i\,(e^{j\omega}; \Delta_k) = \frac{S_i(e^{j\omega}; \Delta_k)}{R_i(e^{j\omega}; \Delta_k)} \tag{9-14}$$

In this way, we are dividing the short-time windowing effects in two categories: those associated with a simple change in amplitudes of the wavelet arrival times, as represented by $r_i[n; \Delta_k]$, and those associated with a distortion in the actual pulse shape, as represented by $w_i[n; \Delta_k]$.

The choice of this model among the infinite variety of ways of expressing $s_i[n; \Delta_k]$ in terms of a convolution of two components is based on the premise that the ultimate goal of the analysis is the recovery of the reflector segment $r[n; \Delta_k]$, which, given the short-time window $g_i[n; \Delta_k]$, simply amounts to the recovery of $r_i[n; \Delta_k]$, a problem that may potentially be handled by the homomorphic analysis of the model of equation (9-13).

Since homomorphic analysis capitalizes on the cepstral properties of the signal components, and the cepstral properties of the impulse train $r_i[n; \Delta_k]$ have already been discussed in Chapter III, we shall now concentrate on the characterization of the cepstral properties of $w_i[n; \Delta_k]$. Such characterization was investigated through a number of experiments. Four of these experiments are described below, all of them along the following pattern. A synthetic time-invariant seismic segment $s[n; \Delta_k]$, resulting from the convolution of a reflector segment $r[n; \Delta_k]$ with a source pulse $w[n; \Delta_k]$ is windowed by windows $g_i[n; \Delta_k]$, $i = 1, 2, 3, 4$. The resulting windowed segments $s_i[n: \Delta_k]$ and their components $r_i[n; \Delta_k]$ and $w_i[n; \Delta_k]$ were computed, as well as their complex cepstra, $\hat{s}_i[n; \Delta_k]$, $\hat{r}_i[n: \Delta_k]$, and $\hat{w}_i[n; \Delta_k]$. Finally, the cepstral analysis of each $\hat{s}_i[n; \Delta_k]$ in terms of its low-time and high-time components yields sequences that will be denoted $s_i^L[n; \Delta_k]$ and $s_i^H[n; \Delta_k]$, respectively. The results are depicted in Figures 9.3 through 9.6. Let us now comment on these experiments.

The first experiment illustrates the effect of a window that converts the reflector series segment into a minimum-phase sequence. The resulting $w_1[n; \Delta_k]$ is a long sequence, characterized by the periodic repetition of its onset, with a period equal to the first interarrival time T_1, of the impulse train $r_1[n; \Delta_k]$. Such periodicity is also evident in the cepstral domain, where a strong correlation is seen between the high-time structures of $\hat{w}_1[n; \Delta_k]$ and $\hat{r}_1[n; \Delta_k]$. The particular structure of the pulse $w_1[n; \Delta_k]$ can be interpreted as follows. The pulse onset, which must be, of course, precisely identical to the segment's onset up to time T_1, has little to do with the underlying pulse shape $w[n; \Delta_k]$. Thus, the

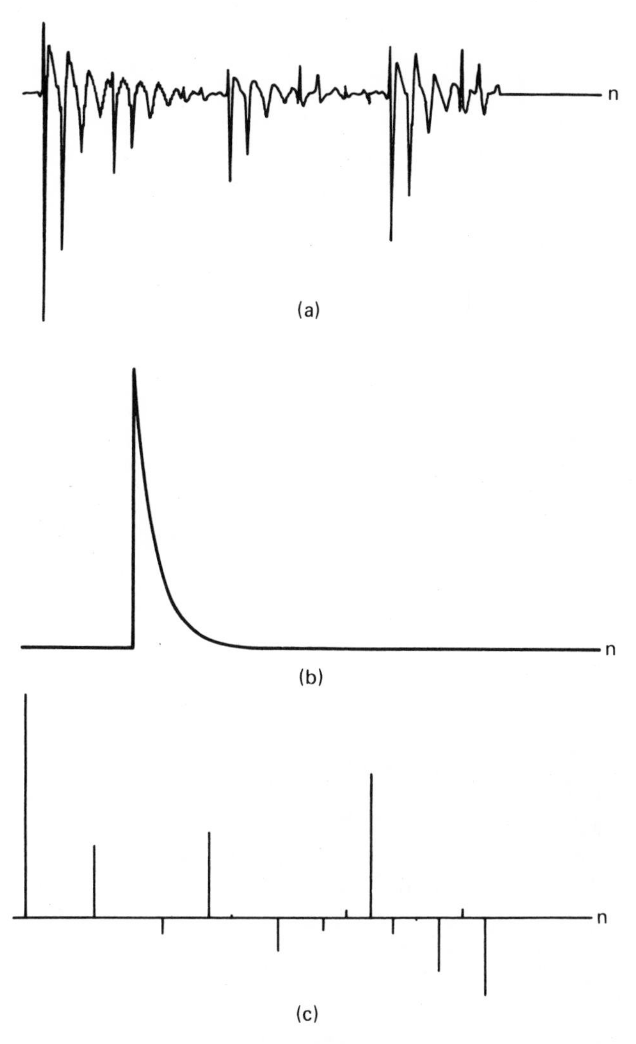

Figure 9.3 Short-time homomorphic analysis using an exponential window: (a) short-time segment

$$s[n;\Delta_k] = [w[n;\Delta_k] \star r[n;\Delta_k]]g[n;\Delta_k];$$

(b) exponential window $g_1[n;\Delta_k]$; (c) seismic reflector segment $r[n;\Delta_k]$

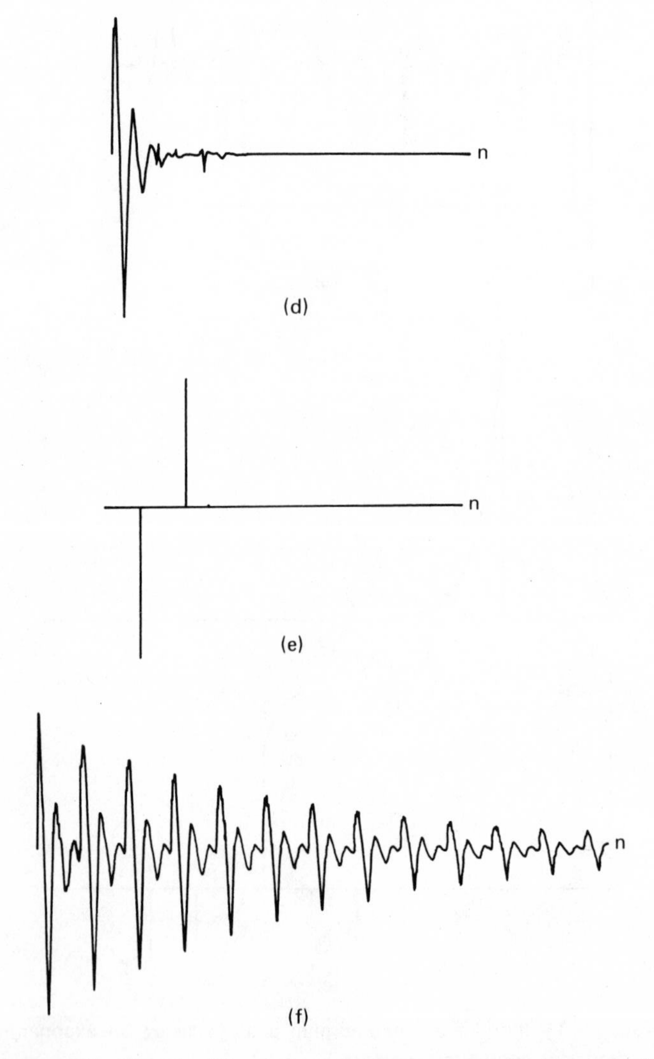

Figure 9.3 (Cont.): (d) windowed seismic segment $s_1[n;\,\Delta_k]$; (e) windowed reflector segment $r_1[n;\,\Delta_k]$; (f) short-time convolutional component $w_1[n;\,\Delta_k]$

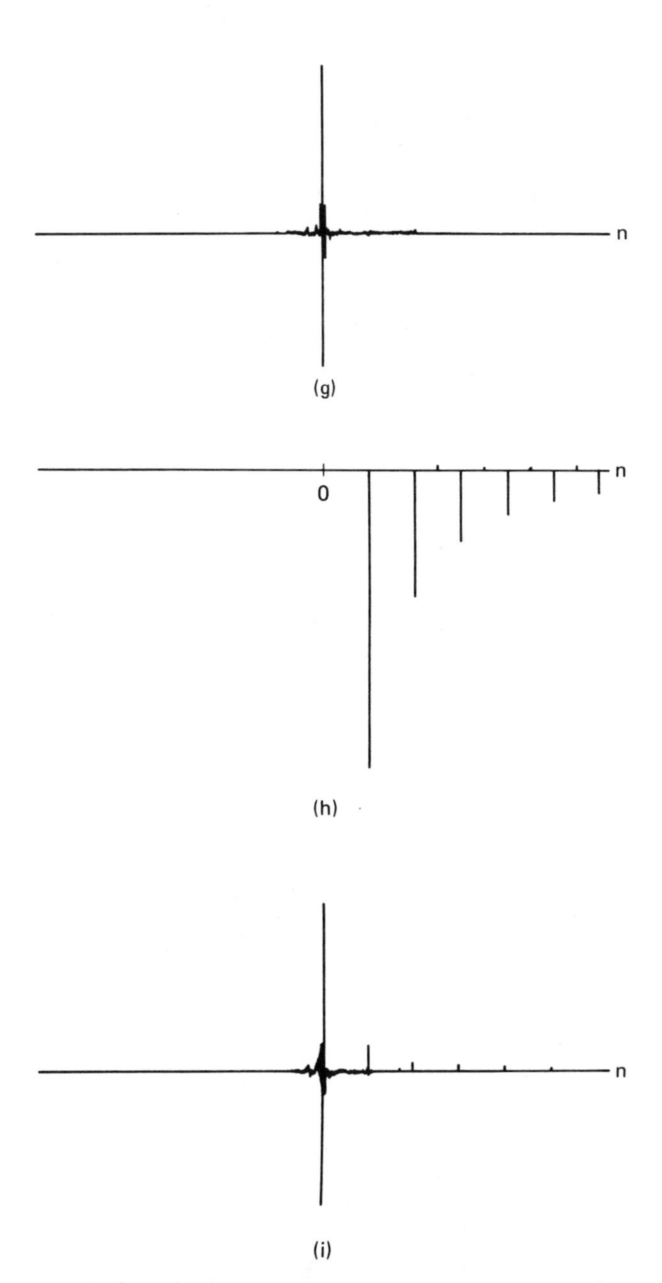

Figure 9.3 (Cont.): (g) complex cepstrum of windowed segment $\hat{s}_1[n;\Delta_k]$; (h) complex cepstrum of windowed reflector segment $\hat{r}_1[n;\Delta_k]$ (the vertical scale has been magnified × 9); (i) complex cepstrum of convolutional component

$$\hat{w}_1[n;\Delta_k] = \hat{s}_1[n;\Delta_k] - \hat{r}_1[n;\Delta_k]$$

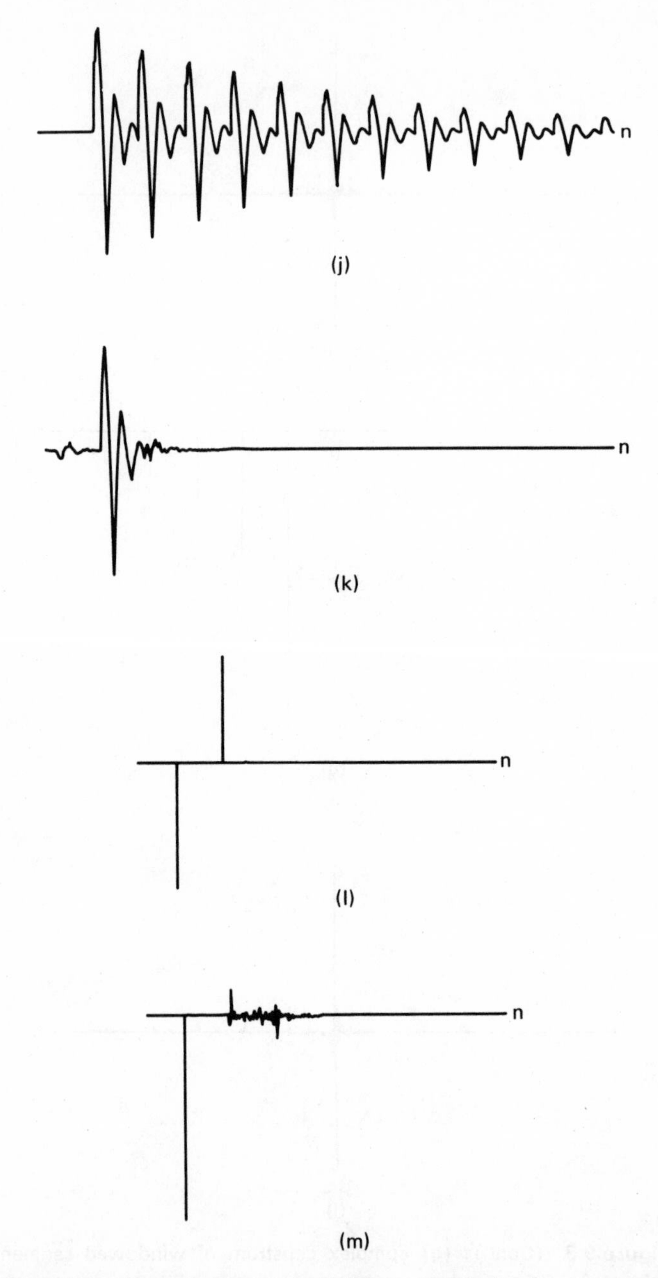

Figure 9.3 (Cont.): (j) short-time convolutional component $w_1[n; \Delta_k]$; (k) low-time cepstral component of windowed segment $s_1^L[n; \Delta_k]$; (l) windowed reflector segment $r_1[n; \Delta_k]$; (m) high-time cepstral component of windowed segment $s_1^H[n; \Delta_k]$

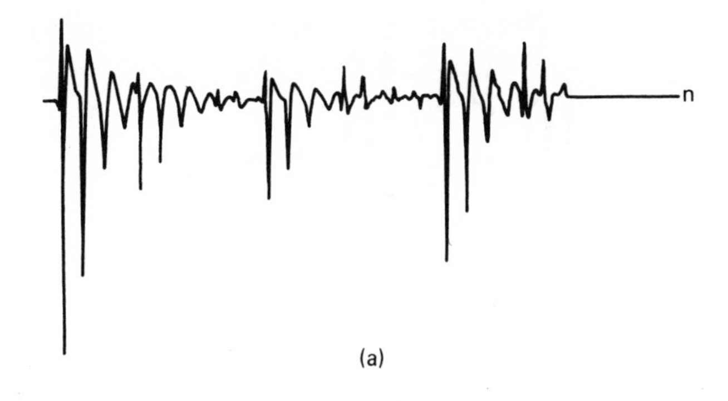

(a)

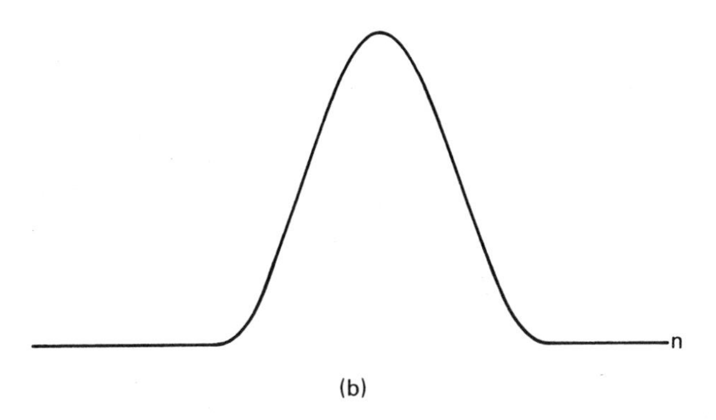

(b)

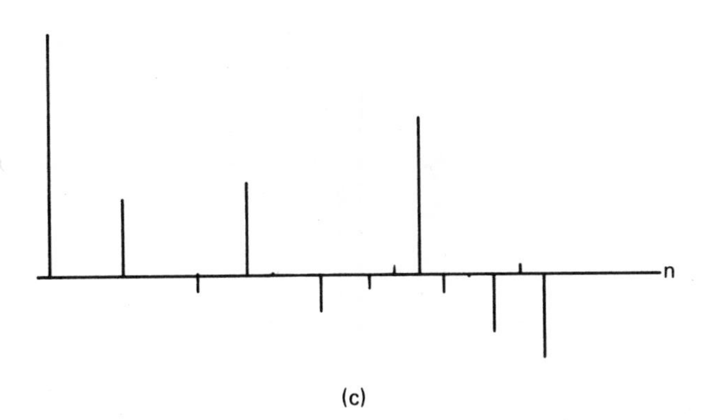

(c)

Figure 9.4 Short-time homomorphic analysis using a Hamming window: (a) short-time segment

$$s[n;\Delta_k] = [w[n;\Delta_k] \star r[n;\Delta_k]]g[n;\Delta_k];$$

(b) Hamming window $g_2[n;\Delta_k]$; (c) seismic reflector segment $r[n;\Delta_k]$

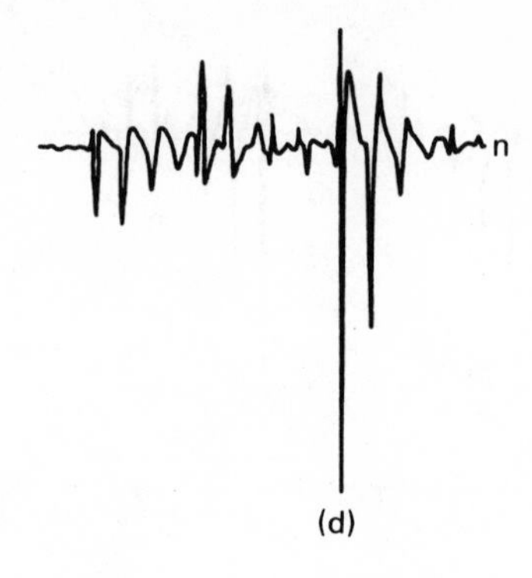

(d)

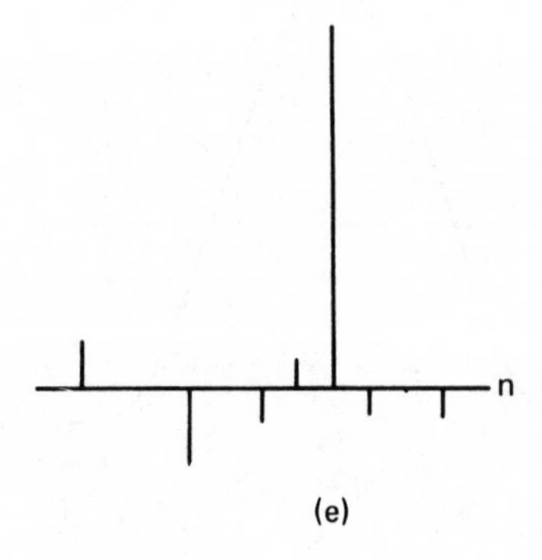

(e)

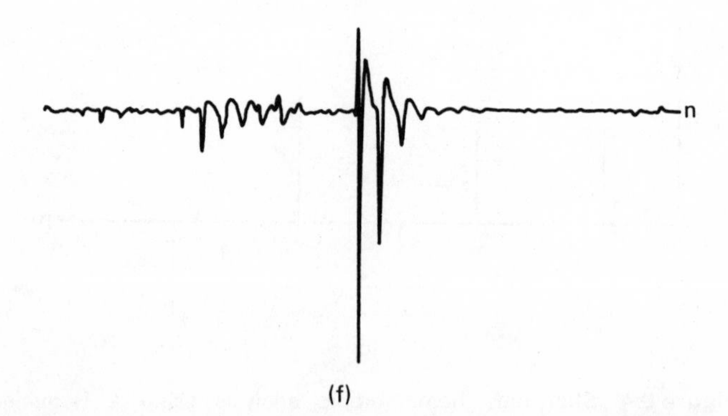

(f)

Figure 9.4 (Cont.): (d) windowed seismic segment $s_2[n; \Delta_k]$; (e) windowed reflector segment $r_2[n; \Delta_k]$; (f) short-time convolutional component $w_2[n; \Delta_k]$

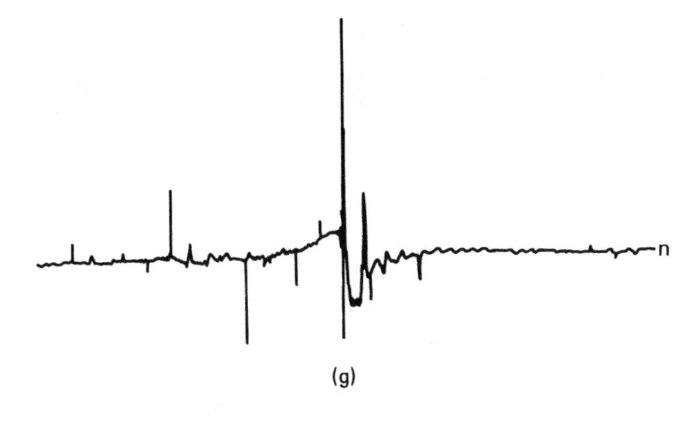

(g)

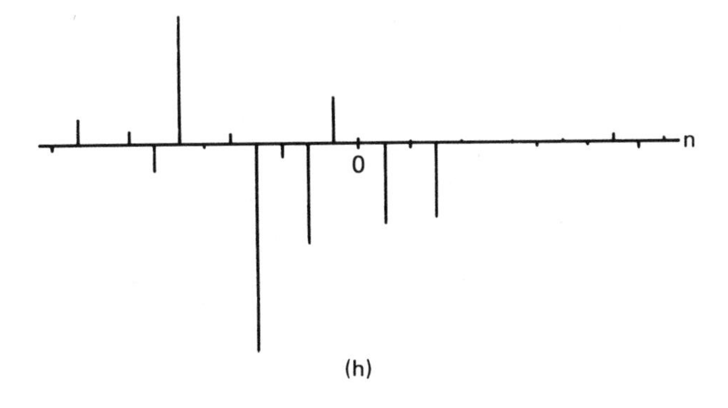

(h)

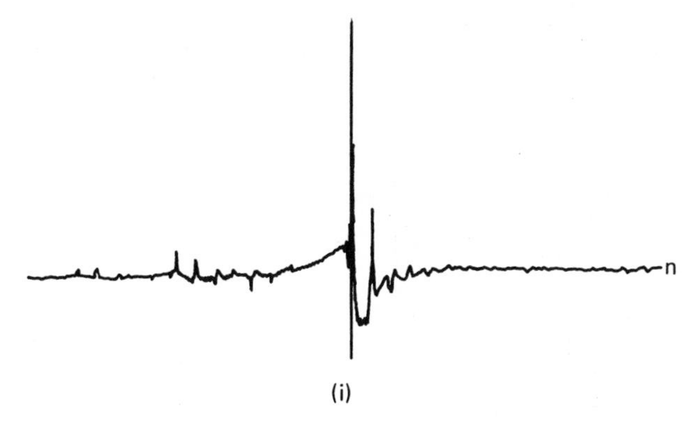

(i)

Figure 9.4 (Cont.): (g) complex cepstrum of windowed segment $\hat{s}_2[n; \Delta_k]$; (h) complex cepstrum of windowed reflector segment $\hat{r}_2[n; \Delta_k]$ (the vertical scale has been magnified × 2.5); (i) complex cepstrum of convolutional component

$$\hat{w}_2[n; \Delta_k] = \hat{s}_2[n; \Delta_k] - \hat{r}_2[n; \Delta_k]$$

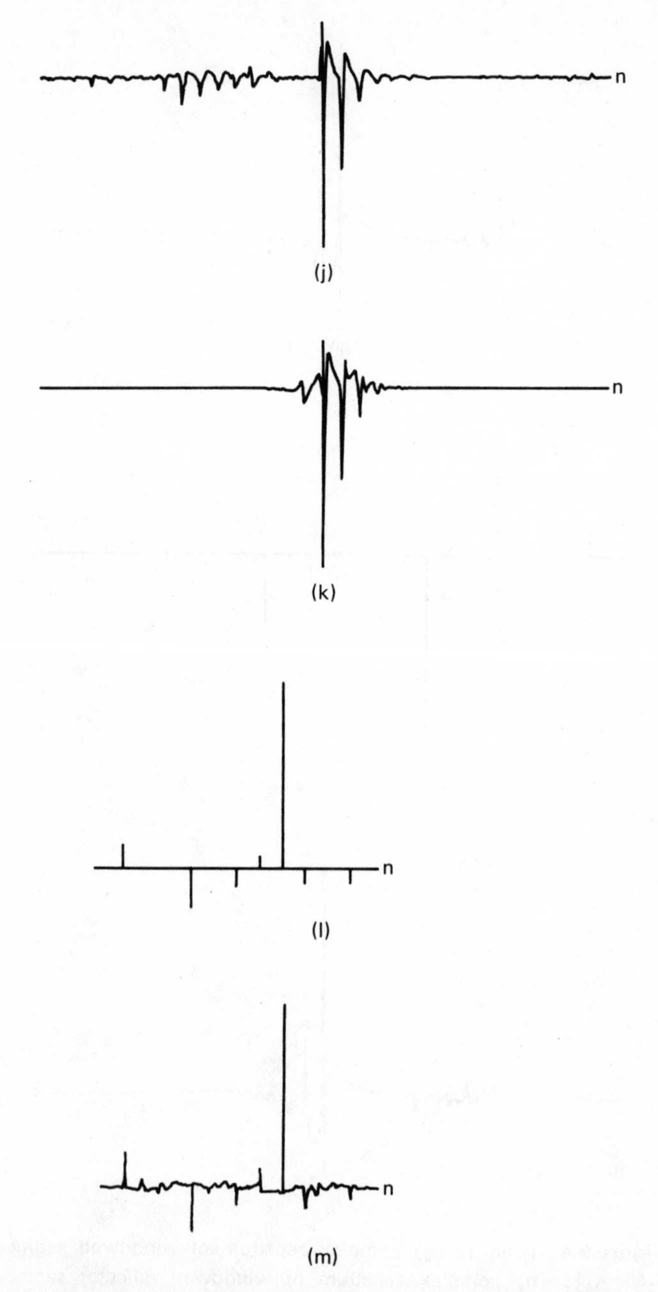

Figure 9.4 (Cont.): (j) short-time convolutional component $w_2[n; \Delta_k]$; (k) low-time cepstral component of windowed segment $s_2^L[n; \Delta_k]$; (l) windowed reflector segment $r_2[n; \Delta_k]$; (m) high-time cepstral component of windowed segment $s_2^H[n; \Delta_k]$

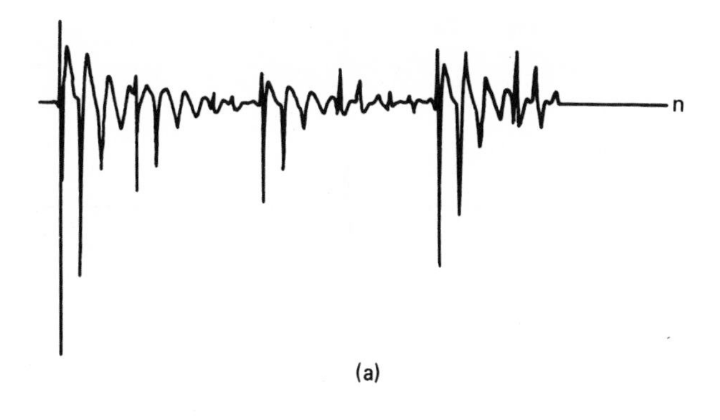

(a)

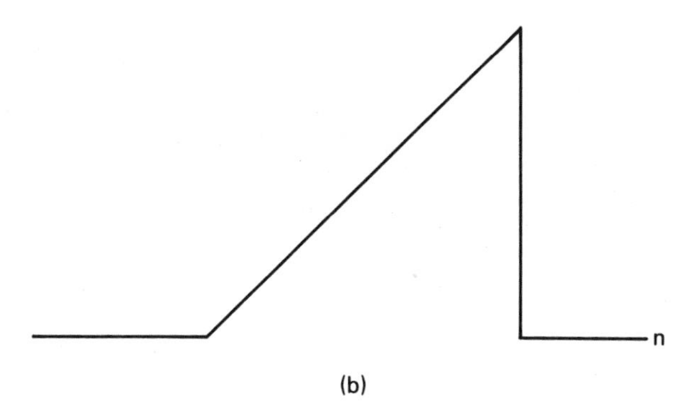

(b)

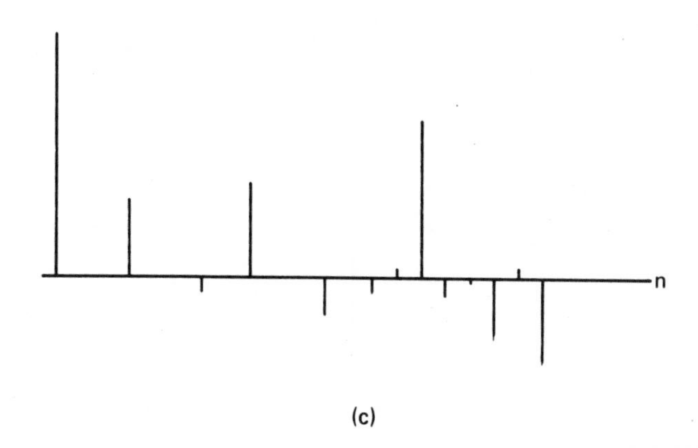

(c)

Figure 9.5 Short-time homomorphic analysis using a linear window: (a) short-time segment $s[n;\Delta_k] = [w[n;\Delta_k] \star r[n;\Delta_k]]g[n;\Delta_k]$; (b) linear window $g_3[n;\Delta_k]$; (c) seismic reflector segment $r[n;\Delta_k]$

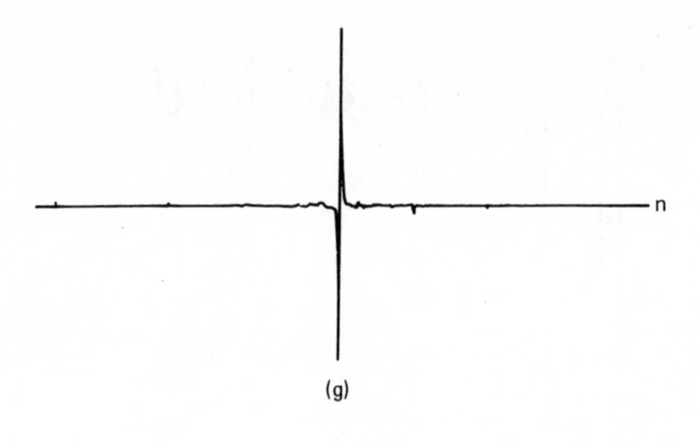

(g)

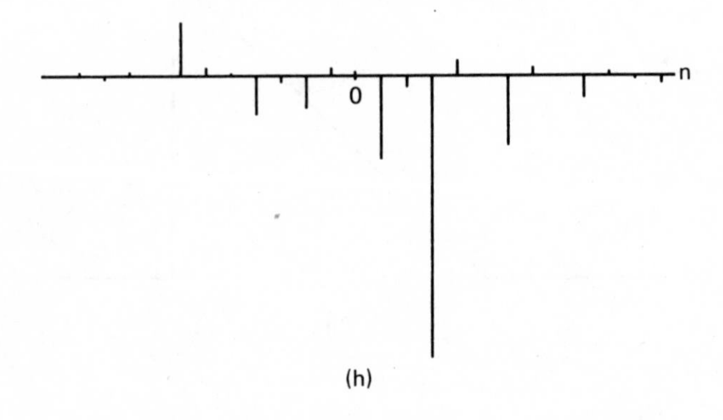

(h)

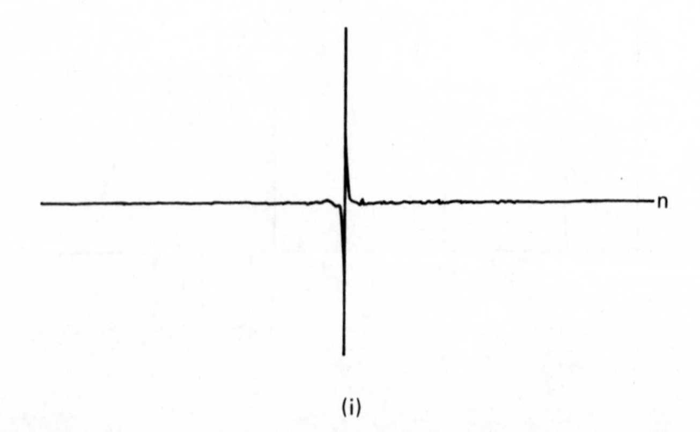

(i)

Figure 9.5 (Cont.): (d) windowed seismic segment $s_3[n; \Delta_k]$; (e) windowed reflector segment $r_3[n; \Delta_k]$; (f) short-time convolutional component $w_3[n; \Delta_k]$

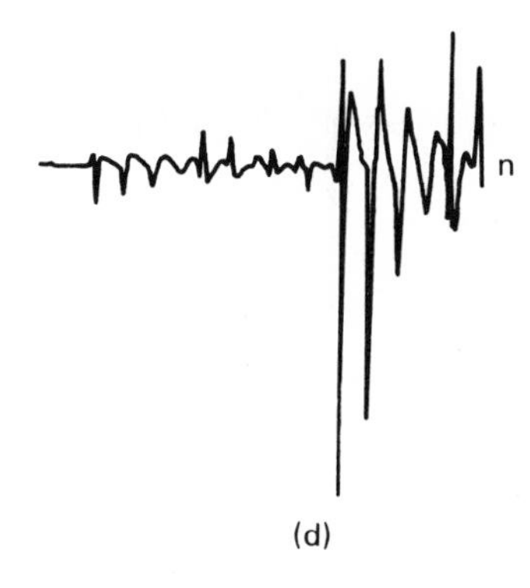

(d)

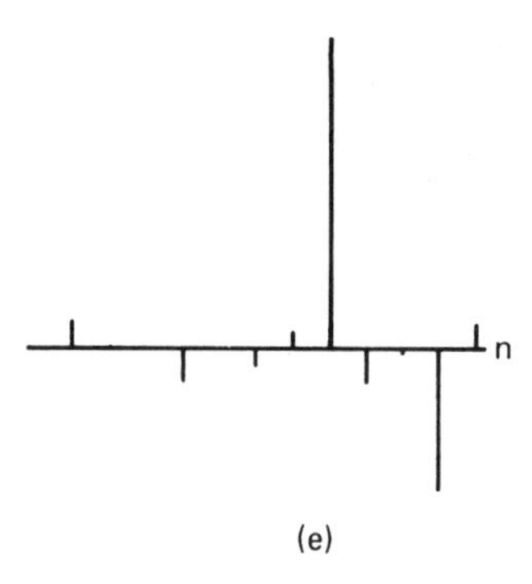

(e)

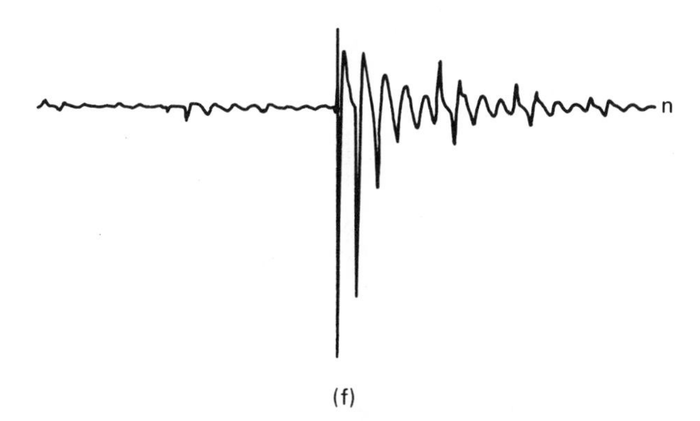

(f)

Figure 9.5 (Cont.): (g) complex cepstrum of windowed segment $\hat{s}_3[n; \Delta_k]$; (h) complex cepstrum of windowed reflector segment $\hat{r}_3[n; \Delta_k]$ (the vertical scale has been magnified × 17); (i) complex cepstrum of convolutional component

$$\hat{w}_3[n; \Delta_k] = \hat{s}_3[n; \Delta_k] - \hat{r}_3[n; \Delta_k]$$

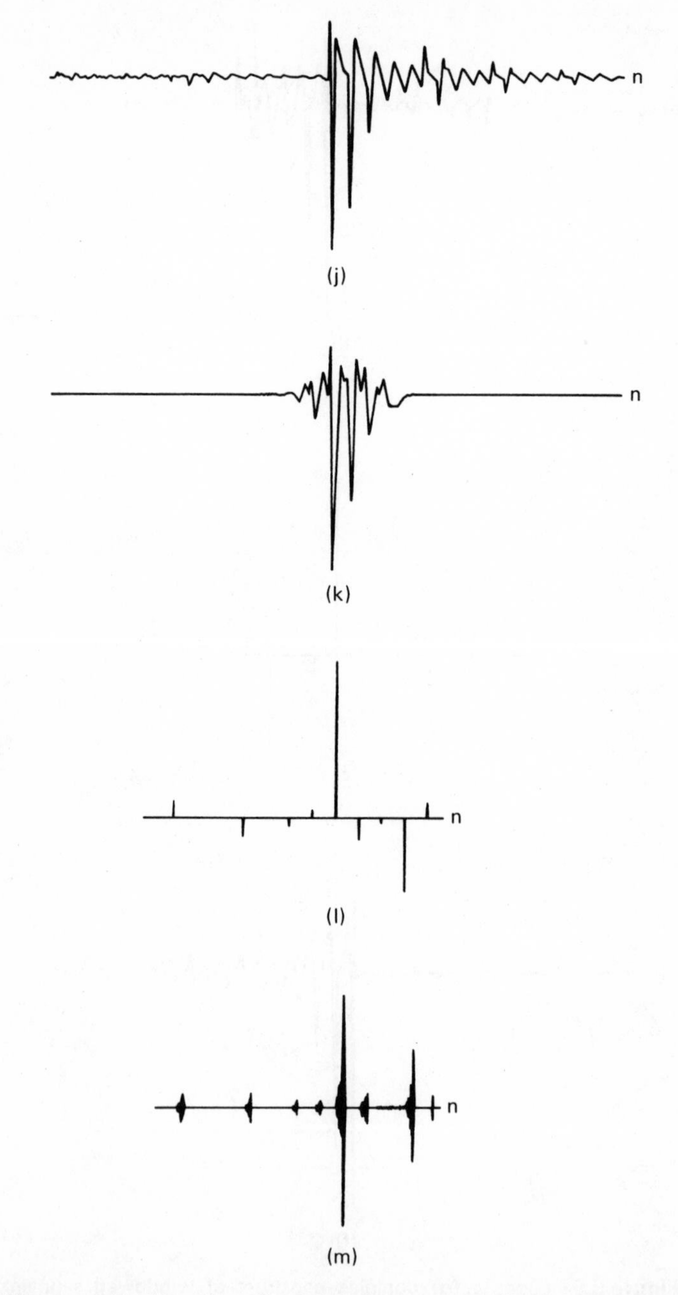

Figure 9.5 (Cont.): (j) short-time convolutional component $w_3[n;\Delta_k]$; (k) low-time cepstral component of windowed segment $s_3^L[n;\Delta_k]$; (l) windowed reflector segment $r_3[n;\Delta_k]$; (m) high-time cepstral component of windowed segment $s_3^H[n;\Delta_k]$

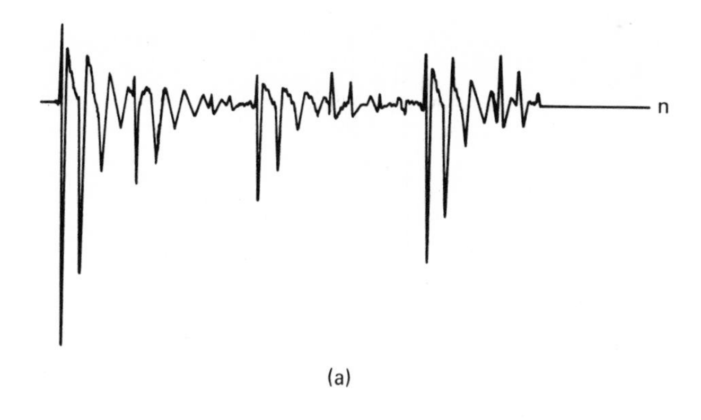

(a)

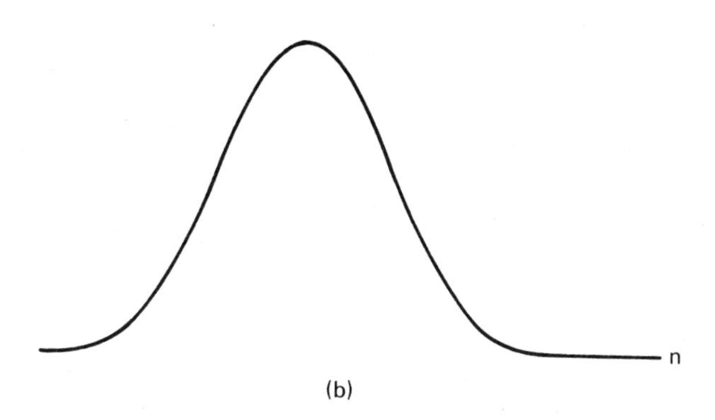

(b)

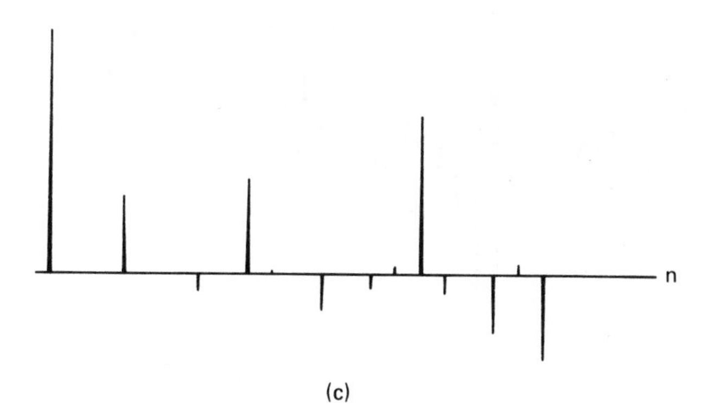

(c)

Figure 9.6 Short-time homomorphic analysis using a Hamming window squared: (a) short-time segment

$$s[n;\Delta_k] = [w[n;\Delta_k] \star r[n;\Delta_k]]g[n;\Delta_k];$$

(b) Hamming window squared $g_4[n;\Delta_k]$; (c) seismic reflector segment $r[n;\Delta_k]$

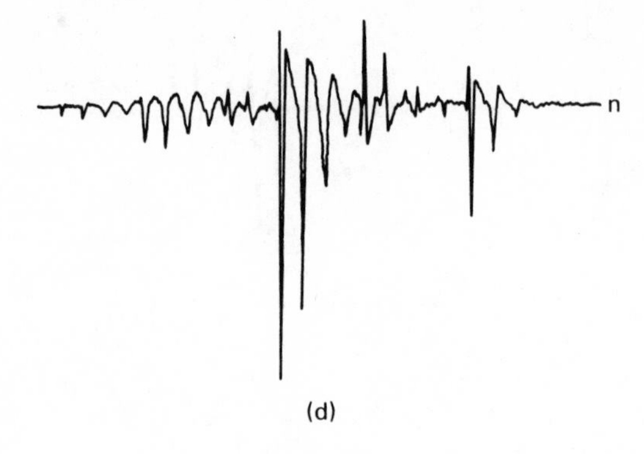

(d)

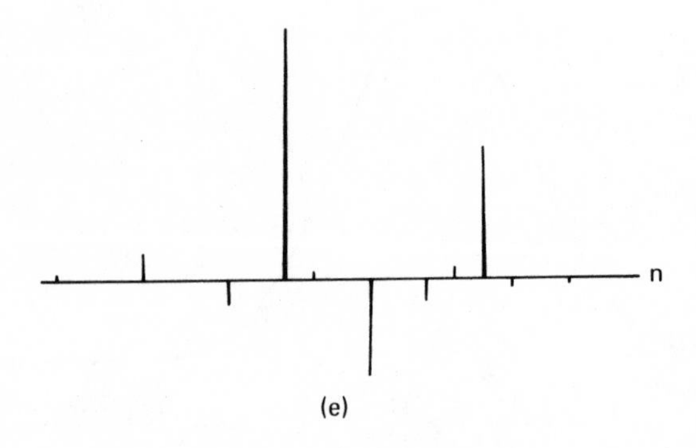

(e)

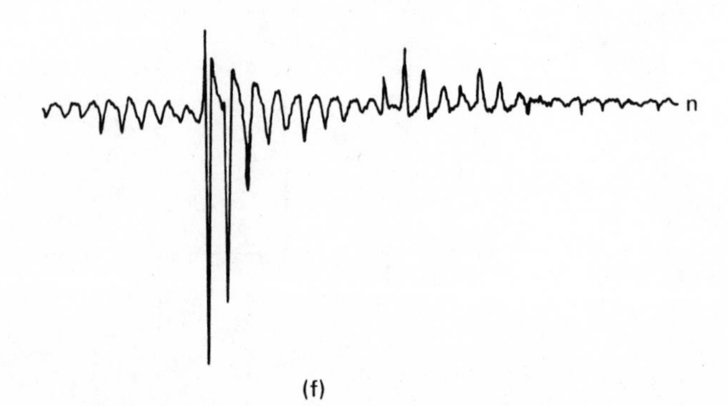

(f)

Figure 9.6 (Cont.): (d) windowed seismic segment $s_4[n; \Delta_k]$; (e) windowed reflector segment $r_4[n; \Delta_k]$; (f) short-time convolutional component $w_4[n; \Delta_k]$

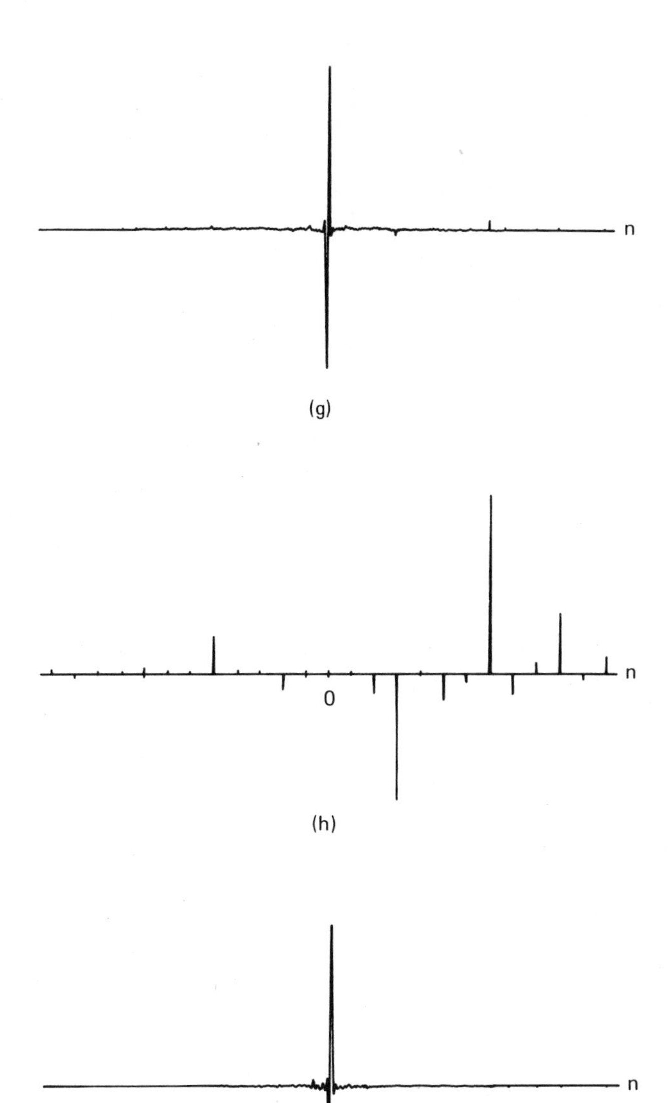

Figure 9.6 (Cont.): (g) complex cepstrum of windowed segment $\hat{s}_4[n; \Delta_k]$; (h) complex cepstrum of windowed reflector segment $\hat{r}_4[n; \Delta_k]$ (the vertical scale has been magnified × 17); (i) complex cepstrum of convolutional component

$$\hat{w}_4[n; \Delta_k] = \hat{s}_4[n; \Delta_k] - \hat{r}_4[n; \Delta_k]$$

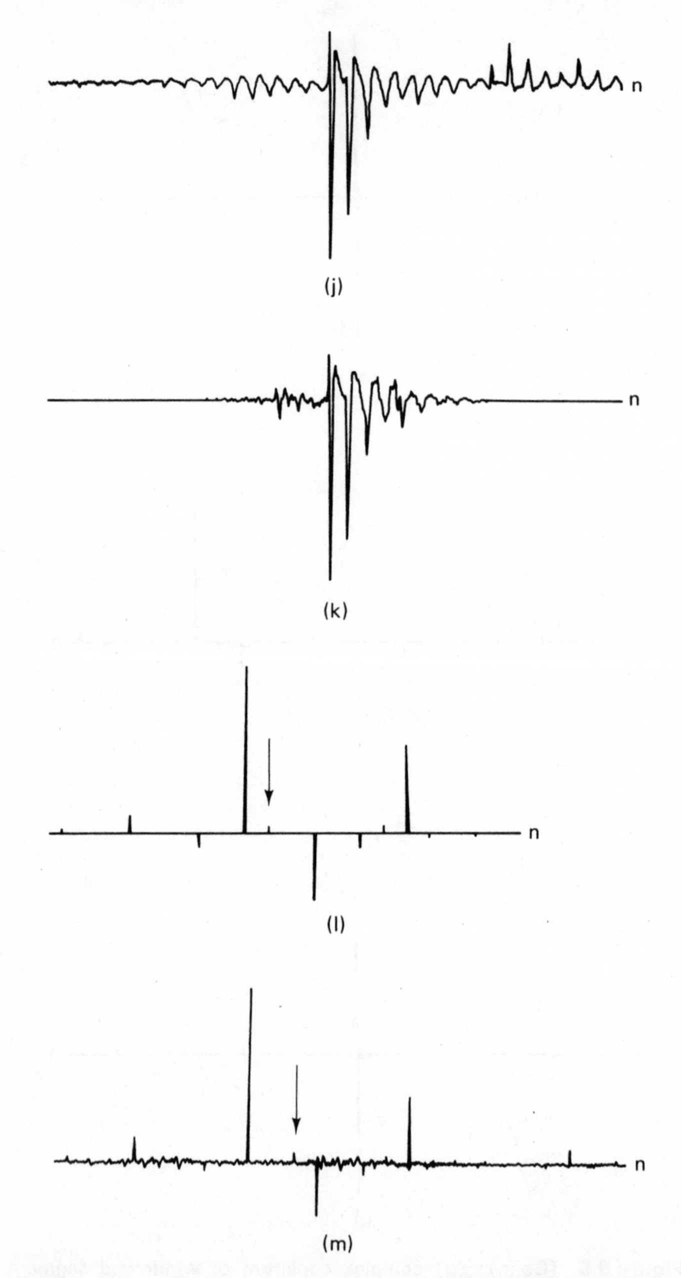

Figure 9.6 (Cont.): (j) short-time convolutional component $w_4[n;\Delta_k]$; (k) low-time cepstral component of windowed segment $s_4^L[n;\Delta_k]$; (l) windowed reflector segment $r_4[n;\Delta_k]$ (arrow indicates misplacement of arrival); (m) high-time cepstral component of windowed segment $s_4^H[n;\Delta_k]$ (arrow indicates misplacement of arrival)

structure of $w_1[n; \Delta_k]$ after T_1 and up to $T_1 + T_2$, where T_2 denotes the second arrival time of the impulse train $r_1[n; \Delta_k]$, corresponds to a correction to the pulse onset, by subtracting it using the second arrival. Since, in this particular example, the windowed reflector segment is dominated by the first two pulses, this same mechanism has to be repeated over and over again, in order to generate the various arrivals of the wavelet $w[n; \Delta_k]$. Therefore, the actual information concerning $w[n; \Delta_k]$ is represented as minor perturbations of the otherwise periodic structure of $w_1[n; \Delta_k]$.

The low-time/high-time cepstral decomposition of $s_1[n; \Delta_k]$ yields a low-time estimate which closely matches the onset of $w_1[n; \Delta_k]$ and therefore the segment's onset, as would be expected from the minimum-phase nature of $r_1[n; \Delta_k]$, as discussed in Chapter VIII. Note, therefore, that the low-time cepstral component of $w_1[n; \Delta_k]$ is very much dependent on the window characteristics, in particular on its onset, and that this component is not associated with the seismic wavelet $w[n; \Delta_k]$. Similar results hold, in general, whenever the short-time window makes the reflector segment minimum-phase or maximum-phase.

The second experiment illustrates the effects of a Hamming window. Now, $r_2[n; \Delta_k]$ is a mixed-phase impulse train, and the resulting $w_2[n; \Delta_k]$ is seen to capture the seismic wavelet $w[n; \Delta_k]$, although the windowing end effects can still be observed, particularly at the segment onset.

The seismic wavelet information is now seen to be represented in the low-time cepstral region, as is demonstrated by $s_2^L[n; \Delta_k]$. The reflector series information is represented in the high-time region, where it overlaps a residual component due to $\hat{w}_2[n; \Delta_k]$, leading to the noisy high-time homomorphic estimate $s_2^H[n; \Delta_k]$. This experiment therefore illustrates the potential of homomorphic systems in analyzing short-time data. By allowing $r_2[n; \Delta_k]$ to be mixed-phase, the onset matching phenomena observed in the first experiment has been avoided, with the resulting improvement in the estimation of the underlying convolutional signal components.

The third experiment illustrates the use of a short-time linear ramp. In this case, $r_3[n; \Delta_k]$ is a mixed-phase impulse train, and $w_3[n; \Delta_k]$ still captures the basic shape of $w[n; \Delta_k]$, although with significant offset disturbances due to the abrupt discontinuity of the window. The low-time homomorphic estimate $s_3^L[n; \Delta_k]$ is still highly correlated with $w[n; \Delta_k]$ and the high-time estimate $s_3^H[n; \Delta_k]$ is seen to correlate with $r_3[n; \Delta_k]$ in terms of the impulse arrival times and their amplitudes except that we are now faced with doublet-type arrivals, thus rendering the actual determination of the impulse polarities rather difficult.

Finally, the fourth experiment deals with a Hamming window squared, with the purpose of illustrating the effects of the distortion of the convolutional model due to sharp variation of the window simultaneously at its offset and onset. This is, in fact, evident in $w_4[n; \Delta_k]$, where the seismic pulse $w[n; \Delta_k]$ is seen, but with significant disturbances at both ends. The reflector segment $r_4[n; \Delta_k]$ is again mixed-phase, and for the first time in this set of experiments is seen to have small contributions in the low-time cepstral region. The high-time estimate $s_4^H[n; \Delta_k]$ is now seen to be corrupted not only by the type of convolutional noise displayed in the previous examples, which is due to the incorporation of the high-time components of $\hat{w}_4[n; \Delta_k]$, but also by a more subtle kind of convolutional noise that is due to the loss of the low-time cepstral components of $\hat{r}_4[n; \Delta_k]$. This noise component is responsible for the misplacement of the impulse indicated by the arrow in Figure 9.6 m and in general by the slight amplitude distortion of the overall impulse train, as illustrated in Figure 9.7, where the upper sequence corresponds to the high-time estimate convolved with the low-time component due to $\hat{r}_4[n; \Delta_k]$.

The set of experiments described above illustrates the basic features to be taken into consideration in the homomorphic analysis of a short-time seismic segment $s_i[n; \Delta_k]$:

1. The use of windows that make $r_i[n; \Delta_k]$ minimum-phase or maximum-phase must be avoided, owing to the corresponding onset and offset matching implications.
2. The use of windows that make $r_i[n; \Delta_k]$ mixed-phase leads to sequences $w_i[n; \Delta_k]$, which capture the seismic wavelet $w[n; \Delta_k]$ in its low-time cepstral region.
3. In general, the cepstra of $r_i[n; \Delta_k]$ and $w_i[n; \Delta_k]$ do overlap. In fact, although the cepstral characteristics of $r_i[n; \Delta_k]$ can be dramatically changed by changing the window charac-

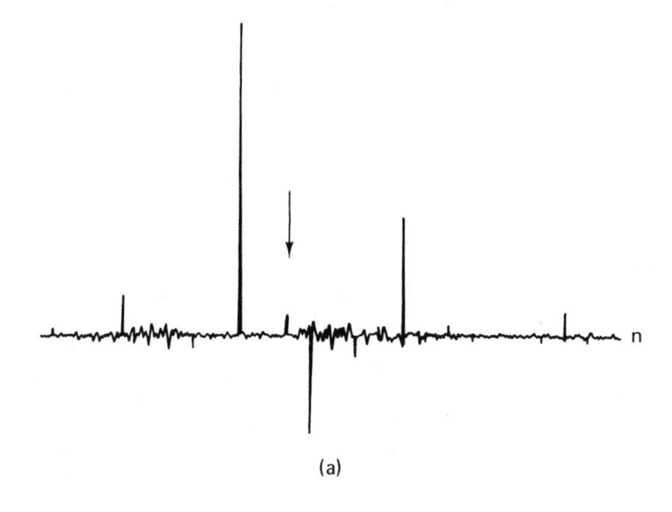

(a)

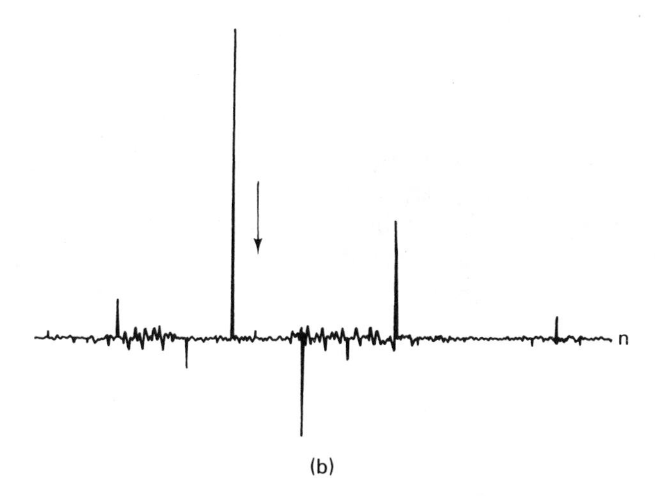

(b)

Figure 9.7 Convolutional noise effects on reflector series estimation by high-time cepstral gating: (a) high-time component of short-time segment windowed with a Hamming window squared $s_4^H[n; \Delta_k]$; (b) high-time component $s_4^H[n; \Delta_k]$ convolved with low-time component of windowed reflector segment $r_4^L[n; \Delta_k]$

teristics, there seems to be, at the present, no apparent way of designing, a priori, a window that would be guaranteed to enforce high-time cepstral characteristics for the mixed-phase reflector segment. Furthermore, the windowing operation will always introduce some amount of distortion, which is, as we saw, associated with the high-time cepstral structure of $w_i[n; \Delta_k]$.

These are the major features to be taken into consideration when assessing the role of homomorphic systems in short-time data analysis, to be discussed in Chapter X.

IX.5 SUMMARY

By taking advantage of the slow-time variation of the seismic wavelet characteristics, short-time signal models are developed, which represent the seismic trace as a series of windowed time-invariant models. The homomorphic analysis of such short-time segments implies an understanding of the distortion effects caused by the windowing operation. It was shown that the use of short-time windows that render the reflector-series segments minimum-phase or maximum-phase must be avoided.

It was also shown that the seismic wavelet is represented in the low-time region of the segment's complex cepstrum, as long as the windowed reflector segment has mixed-phase characteristics.

X

Short-time Homomorphic Analysis: A Strategy for Wavelet Estimation

X.1 INTRODUCTION

We shall now investigate strategies for short-time homomorphic seismic data analysis. Let us first quickly review the framework within which this analysis is to take place.

First, we are modeling a seismic segment within an interval Δ_k in terms of short-time windowed time-invariant models of the form

$$s[n; \Delta_k] = [w[n; \Delta_k] \star r[n; \Delta_k]]g[n; \Delta_k] \qquad (10\text{-}1)$$

Second, we concluded that the use of smooth windows $g_i[n; \Delta_k]$, leading to mixed-phase windowed reflector segments $r_i[n; \Delta_k]$, enabled the complex cepstrum of the data to capture, in its low-time region, the information associated with the seismic wavelet

$w[n; \Delta_k]$. More specifically, letting

$$s_i[n; \Delta_k] = s[n; \Delta_k]g_i[n; \Delta_k] = w_i[n; \Delta_k] \star r_i[n; \Delta_k] \tag{10-2}$$

we concluded that the low-time cepstral components of $w[n; \Delta_k]$ correspond to the low-time cepstral components of $w_i[n; \Delta_k]$, that is,

$$\hat{w}^L[n; \Delta_k] \simeq \hat{w}_i^L[n; \Delta_k] \tag{10-3}$$

We recall that we are implicitly assuming the use of bandpass mapping, as required. Thus, the seismic wavelet cepstrum $\hat{w}[n; \Delta_k]$ is now essentially contained in the low-time region, since the major source of high-time components, the out-of-band spectral zeros, have been removed as a consequence of the bandpass mapping.

The impulse train $r_i[n; \Delta_k]$, being mixed-phase, will have, in general, both low-time and high-time cepstral components. Therefore, we are faced with the task of designing a strategy for homomorphically filtering $s_i[n; \Delta_k]$, in the presence of cepstral overlap between the signal components. Given the fact that the seismic wavelet information is essentially contained in the low-time region, we shall restrict ourselves to the use of low-time and high-time cepstral gates. Therefore, the low-time/high-time homomorphic analysis of $s_i[n; \Delta_k]$ yields

$$s_i^L[n; \Delta_k] = w_i^L[n; \Delta_k] \star r_i^L[n; \Delta_k] \tag{10-4a}$$

$$s_i^H[n; \Delta_k] = w_i^H[n; \Delta_k] \star r_i^H[n; \Delta_k] \tag{10-4b}$$

The performance of any analysis strategy involving such homomorphic filtering operations will then ultimately depend on the time-domain properties of $r_i^L[n; \Delta_k]$ and $r_i^H[n; \Delta_k]$.

X.2 HOMOMORPHIC ANALYSIS OF MIXED-PHASE RANDOM IMPULSE TRAINS

We shall now explore the sensitivity of the cepstral structure of a mixed-phase random impulse train to time-domain amplitude perturbations in an attempt to characterize the time behavior of its low-time and high-time components. Toward this goal, a number of experiments were performed, as described below.

Consider the seismic reflector series $r[n]$ of Figure 10.1a and

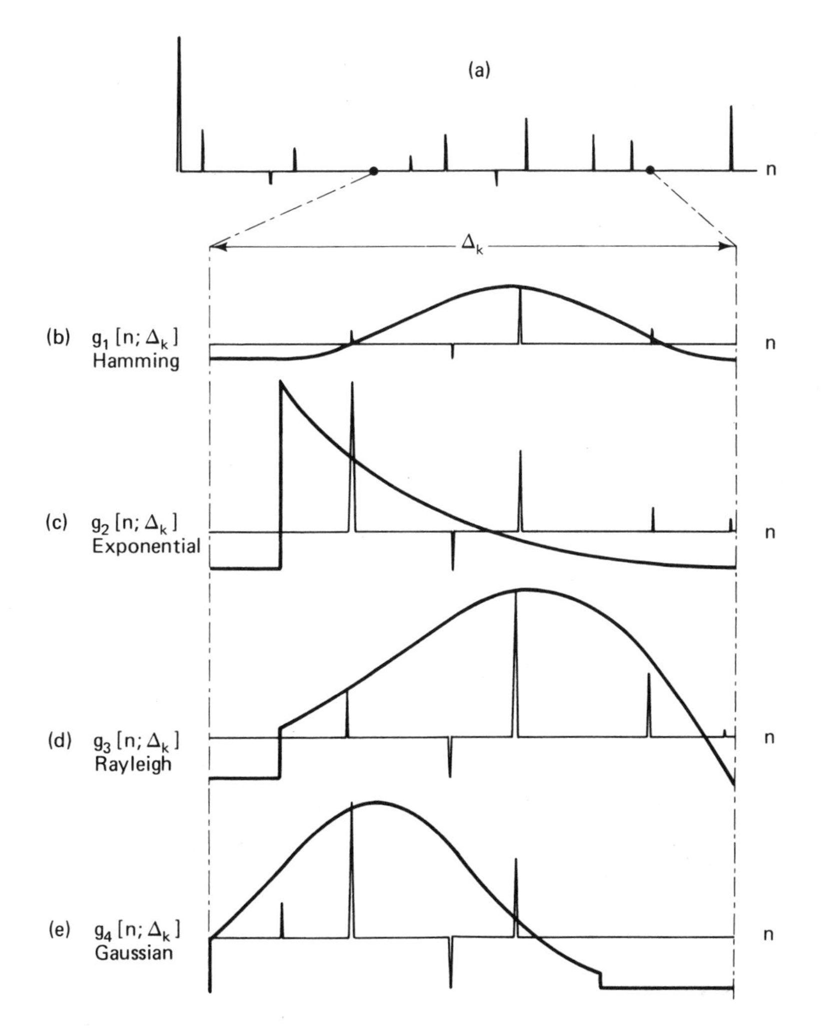

Figure 10.1 Analysis of mixed-phase impulse trains: (a) seismic reflector series $r[n]$; (b)–(e) windows $g_i[n; \Delta_k]$ and windowed reflector segments $r_i[n; \Delta_k]$

the four short-time reflector segments $r_i[n; \Delta_k]$, $i = 1, \ldots, 4$, illustrated in Figures 10.1b through e. These segments result from short-time windowing $r[n]$ with, respectively, a Hamming, an exponential, a Rayleigh, and a Gaussian window, which are also depicted in Figure 10.1. The underlying earth structure is that of Table 10.1.

TABLE 10.1

layer number	*two-way travel time within the layer (ms)*	*reflection coefficient at top interface*
1	111	0.30
2	302	0.10
3	100	−0.03
4	503	0.05
5	154	0.03
6	219	0.08
7	138	−0.04
8	286	0.12
9	167	0.08
10	431	0.06
11	∞	0.15

Figures 10.2 through 10.5 depict, for each segment $r_i[n; \Delta_k]$, the corresponding autocorrelation function $R_{r_i}[n; \Delta_k]$, and the even and odd cepstral components $\hat{r}_{i_e}[n; \Delta_k]$ and $\hat{r}_{i_o}[n; \Delta_k]$. We observe, in all cases, a strong structural correlation between the ACF and the even cepstral components, which is not surprising, since

$$\hat{r}_{i_e}[n; \Delta_k] = \text{IFT}(\tfrac{1}{2} \log (\text{FT}(R_{r_i}[n; \Delta_k]))) \qquad (10\text{-}5)$$

In particular, we note that all these segments are uncorrelated up to lag T_u, shown in the figures, and that, in fact, most of the even cepstral components are essentially confined to the high-time interval $|n| > T_u$.

The odd cepstral component may or may not exhibit a structural correlation with the segment's ACF. In fact, such correlation is seen to exist in Figures 10.2 and 10.4. On the other hand, no such relationship is found in Figures 10.3 and 10.5.

The existence or nonexistence of such a structural link between the ACF and the phase cepstra seems to be associated with the segment's lag, s_{r_i}. In fact, in the course of this research we came across the empirical but consistent observation that, whenever the signal lag equals one of the impulse arrival times, as happens

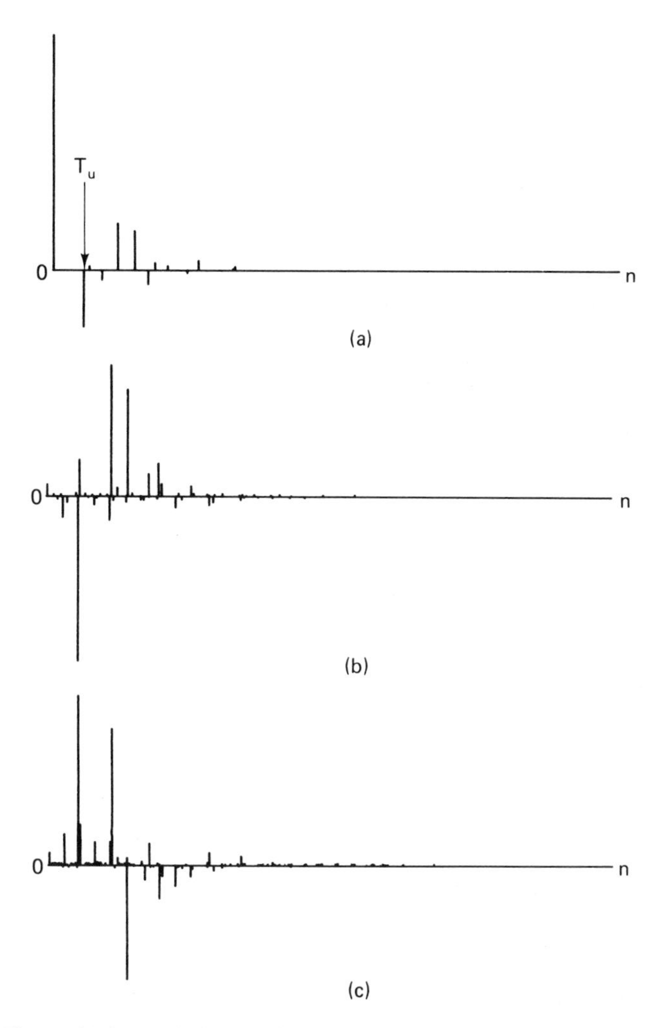

Figure 10.2 Analysis of Hamming windowed reflector segment: (a) autocorrelation function $R_{r_1}[n; \Delta_k]$; (b) even cepstrum $\hat{r}_{1_e}[n; \Delta_k]$; (c) odd cepstrum $\hat{r}_{1_o}[n; \Delta_k]$

in the first and third cases, the link referred to above is indeed in effect. This relationship can be theoretically demonstrated for minimum-phase, maximum-phase, and periodic impulse trains. Whether or not there is a general relationship between these two phenomena is still an open theoretical question.

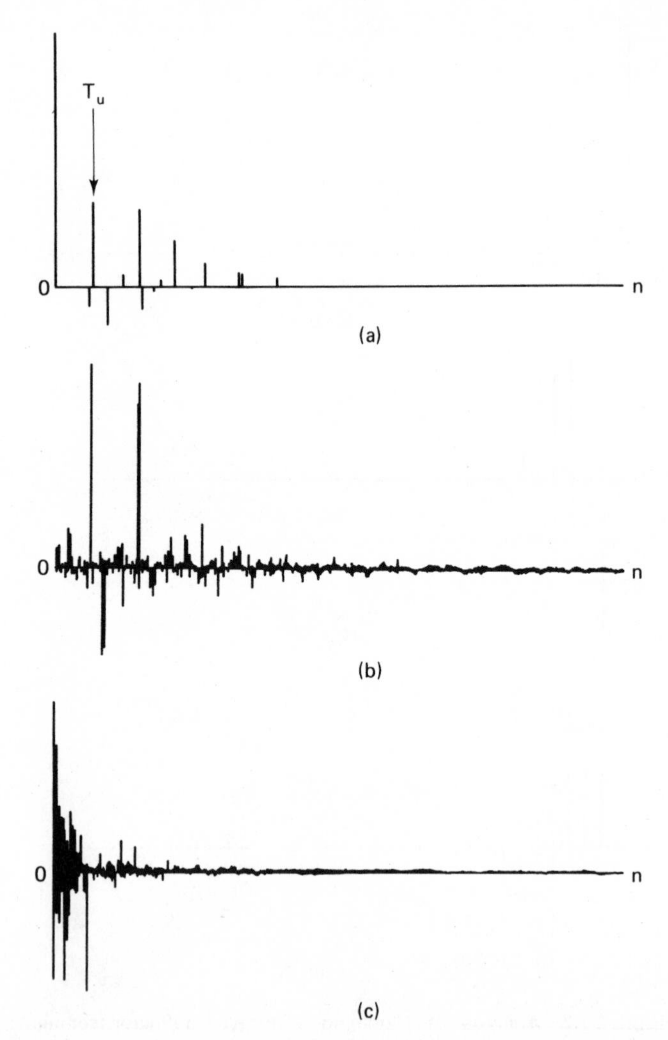

Figure 10.3 Analysis of exponentially windowed reflector segment: (a) autocorrelation function $R_{r_2}[n; \Delta_k]$; (b) even cepstrum $\hat{r}_{2_e}[n; \Delta_k]$; (c) odd cepstrum $\hat{r}_{2_o}[n; \Delta_k]$

In the absence of a better criterion, we shall henceforth define the low-time/high-time cepstral decomposition of $r_i[n; \Delta_k]$ in terms of a cepstral cutoff time that is essentially determined by the uncorrelatedness of the impulse train; that is, we shall define $T_c = T_u$.

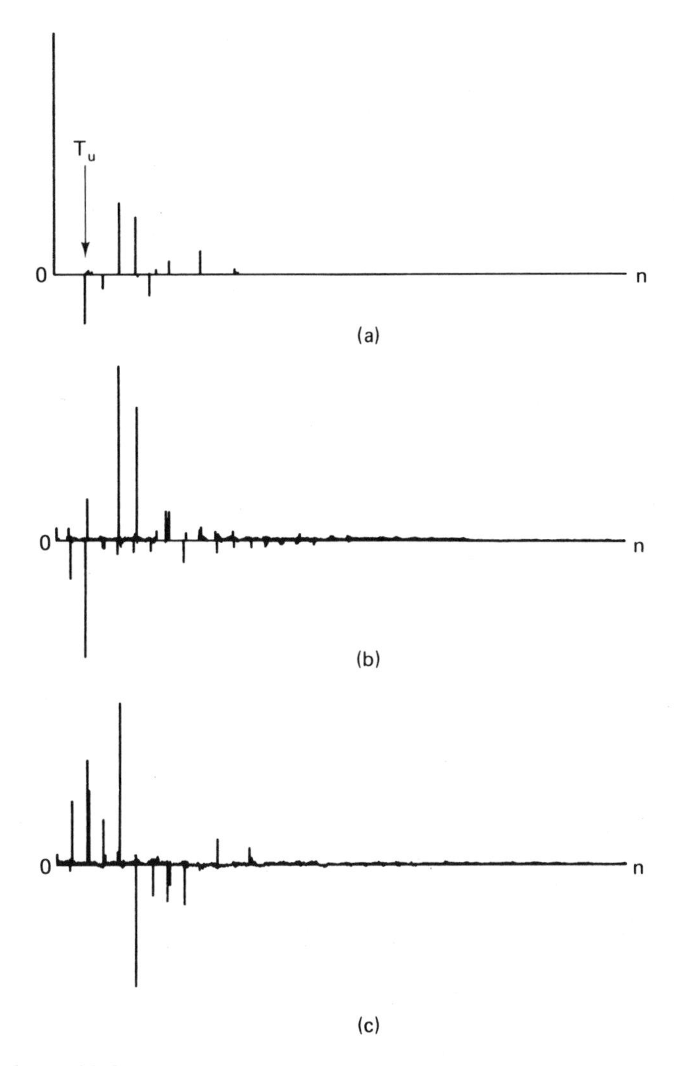

Figure 10.4 Analysis of Rayleigh windowed reflector segment: (a) autocorrelation function $R_{r_3}[n; \Delta_k]$; (b) even cepstrum $\hat{r}_{3_e}[n; \Delta_k]$; (c) odd cepstrum $\hat{r}_{3_o}[n; \Delta_k]$

Figure 10.6 depicts the resulting low-time components $r_i^L[n; \Delta_k]$ after being subject to different synchronization shifts m_i, $i = 1, \ldots, 4$. Let us now attempt to characterize such low-time components in terms of their deviation from the ideal low-time

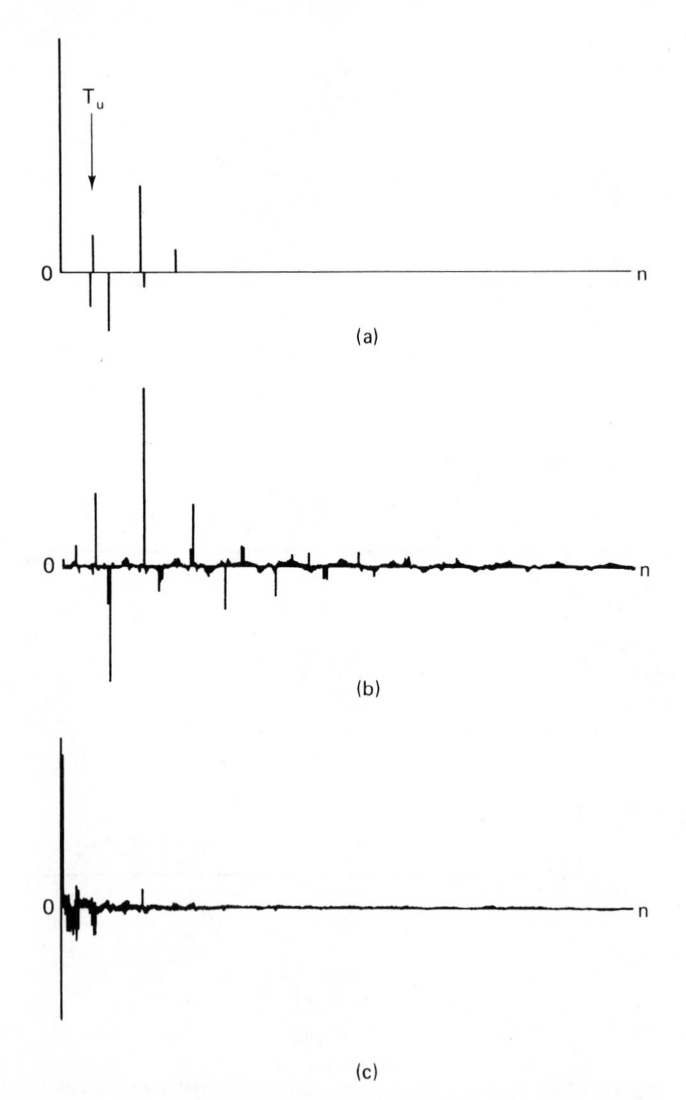

Figure 10.5 Analysis of Gaussian windowed reflector segment: (a) autocorrelation function $R_{r_4}[n; \Delta_k]$; (b) even cepstrum $\hat{r}_{4_e}[n; \Delta_k]$; (c) odd cepstrum $\hat{r}_{4_o}[n; \Delta_k]$

component $\delta[n]$ of a purely high-time mixed-phase impulse train. In general, we may write

$$r_i^L[n; \Delta_k] = \delta[n - m_i] + \eta_i[n] \qquad (10\text{-}6)$$

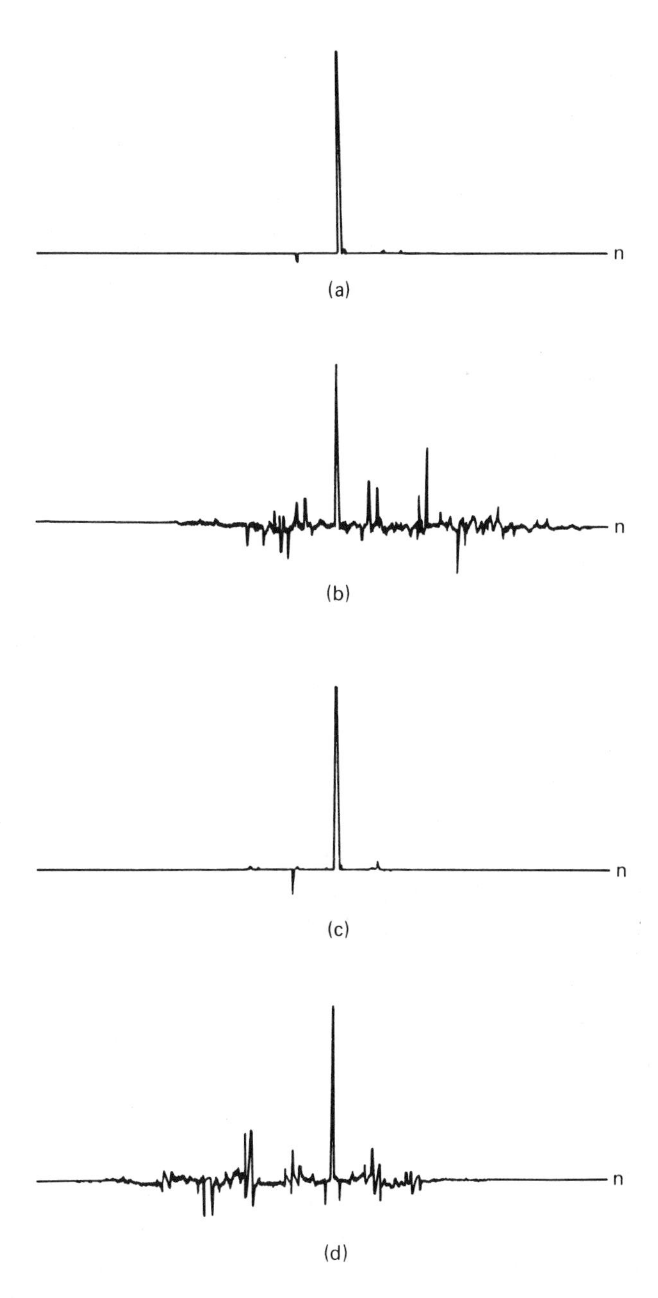

Figure 10.6 Low-time homomorphic filtering of windowed reflector segments: (a) low-time component of Hamming windowed segment $r_1^L[n + m_1; \Delta_k]$; (b) low-time component of exponentially windowed segment $r_2^L[n + m_2; \Delta_k]$; (c) low-time component of Rayleigh windowed segment $r_3^L[n + m_3; \Delta_k]$; (d) low-time component of Gaussian windowed segment $r_4^L[n + m_4; \Delta_k]$

where $\eta_i[n]$ denotes a noise component that may range from almost negligible amplitude levels, as in the first and third cases, to more serious perturbations, as in the other two. We thus observe that the cepstral sensitivity of a mixed-phase impulse train to time-domain amplitude changes is mapped, through the low-time filtering operation, into a similar sensitivity regarding the structure of the noise component $\eta_i[n]$. In other words, it seems reasonable to model the noise component at every instant n as a zero mean random variable with an unknown probability distribution. Thus, defining the *time-stacked* low-time component $\langle r^L[n; \Delta_k]\rangle$ as the average of the various low-time components after proper synchronization, that is,

$$\langle r^L[n; \Delta_k]\rangle = \frac{1}{M}\sum_{i=1}^{M} r_i^L[n + m_i; \Delta_k] = \delta[n] + \langle \eta^L[n]\rangle \tag{10-7}$$

we expect the time-stacked noise component $\langle \eta^L[n]\rangle$ to converge in the limit to its ensemble average.

We illustrate the effects of such averaging technique on the low-time estimates of Figure 10.6. The corresponding stacked sequence is depicted in Figure 10.7a, where a considerable signal-to-noise improvement is readily observable. A similar attitude was discussed in Chapter VIII, involving a stacking procedure in the cepstral domain. We illustrate this approach by depicting in Figure 10.7b the low-time component of the cepstral average $\langle r[n; \Delta_k]\rangle^L$, defined as

$$\langle r[n; \Delta_k]\rangle^L = D_{\star}^{-1}[\langle \hat{r}_i^L[n; \Delta_k]\rangle] \tag{10-8}$$

This technique is based on the assumption that each low-time component $\hat{r}_i^L[n; \Delta_k]$ can be modeled as a zero mean random series. Thus, in the limit, as $M \longrightarrow \infty$, $\langle \hat{r}_i^L[n; \Delta_k]\rangle \longrightarrow 0$, and therefore, in the limit, it will yield the desired $\delta[n]$ in the time domain.

The statistical comparison of time averaging versus (real) logarithmic averaging has received considerable attention in the literature [46]. In contrast, the complex cepstral stacking method does not seem to be amenable to detailed statistical analysis,

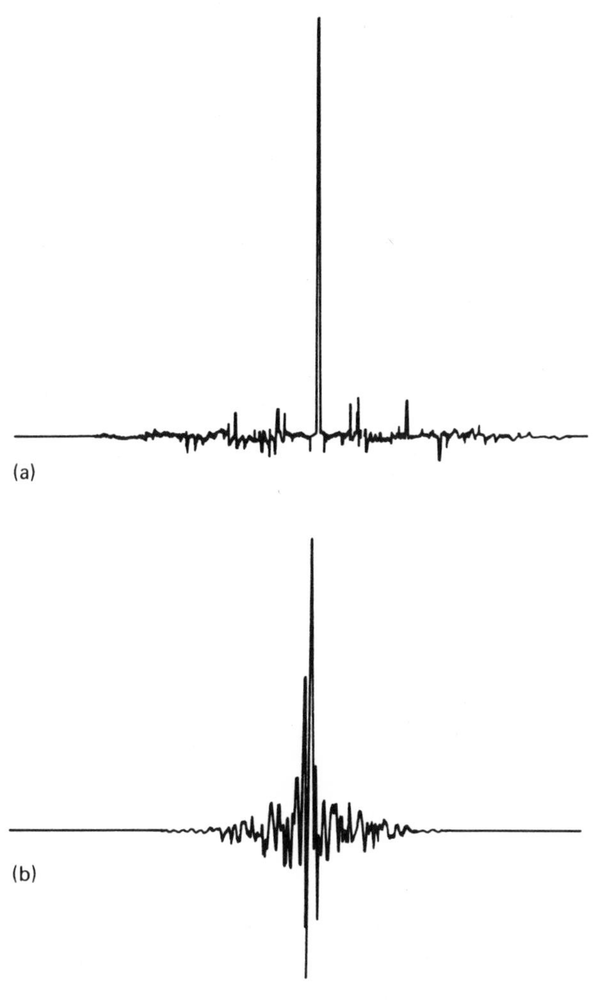

Figure 10.7 Stacking of low-time components: (a) time-stacked low-time component $<r^L[n; \Delta_k]>$; (b) cepstral-stacked low-time component $<r[n; \Delta_k]>^L$

owing to the difficulty in establishing meaningful probabilistic models.

We must thus rely on simulation results, such as those illustrated in Figure 10.7, in attempting to evaluate the performance of this technique. Within this framework, there seems to be enough qualitative difference between the time-stacked and the cepstral-stacked low-time reflector-series components to indicate that this latter technique, if used at all, must be carefully monitored in terms of its convergence behavior.

Let us conclude this section by commenting on the use of high-time homomorphic filters in the recovery of the reflector-

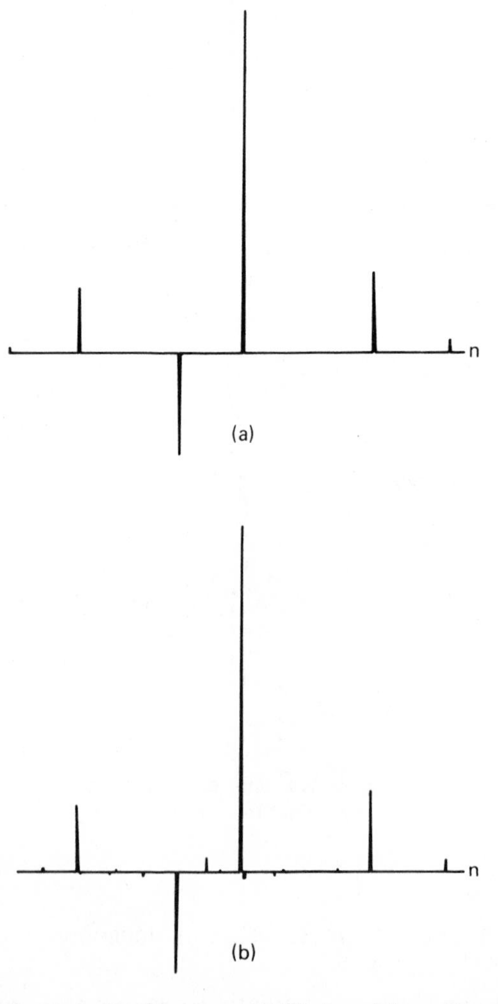

Figure 10.8 High-time homomorphic filtering of Hamming windowed reflector segments: (a) reflector segment $r_1[n; \Delta_k]$; (b) high-time component $r_1^H[n; \Delta_k]$

series segments $r_i[n; \Delta_k]$. Figures 10.8 and 10.9 illustrate the high-time components $r_1^H[n; \Delta_k]$ and $r_4^H[n; \Delta_k]$. The former is clearly a good estimate of $r_1[n; \Delta_k]$, matching indeed the clean impulsive structure of $r_1^L[n; \Delta_k]$. The latter shows, however, very poor correlation with the corresponding segment $r_4[n; \Delta_k]$. In other words, the loss of the low-time cepstral components of a reflector series

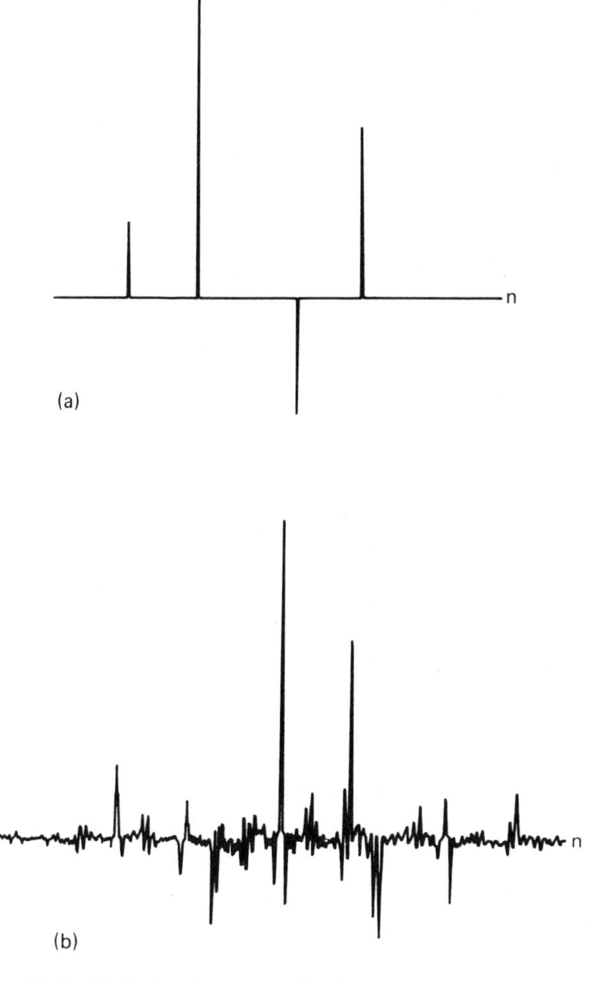

Figure 10.9 High-time homomorphic filtering of Gaussian windowed reflector segment: (a) reflector segment $r_4[n; \Delta_k]$; (b) high-time component $r_1^H[n; \Delta_k]$

may introduce serious convolutional noise, which displays a strong correlation with the data and often leads to very erroneous reflector-series estimates by high-time filtering. Such an approach must then be condemned within this analysis context.

X.3 STRATEGY FOR HOMOMORPHIC WAVELET ESTIMATION

A strategy for short-time homomorphic wavelet estimation may now be outlined, based on the stacking procedure described above. The method consists essentially in performing on a seismic segment $s[n; \Delta_k]$ a number of windowing operations, using a variety of windows $g_i[n; \Delta_k]$, with different shapes, onsets, and lengths. After homomorphic low-time filtering of each of the resulting short-time windowed segments $s_i[n; \Delta_k]$, the resulting outputs $s_i^L[n; \Delta_k]$ are properly synchronized and stacked, yielding the desired wavelet estimate. Consider, then, a set of windowed segments $s_i[n; \Delta_k]$, as in equation (10-2), that is,

$$s_i[n; \Delta_k] = w_i[n; \Delta_k] \star r_i[n; \Delta_k], \qquad i = 1, \ldots, M \tag{10-9}$$

where

$$r_i[n; \Delta_k] = (r[n; \Delta_k]g_i[n; \Delta_k])g[n; \Delta_k] \tag{10-10}$$

Thus, it follows that

$$\hat{s}_i^L[n; \Delta_k] = \hat{w}_i^L[n; \Delta_k] + \hat{r}_i^L[n; \Delta_k], \qquad i = 1, \ldots, M \tag{10-11}$$

and it follows from equation (10-3) that

$$\hat{s}_i^L[n; \Delta_k] = \hat{w}^L[n; \Delta_k] + \hat{r}_i^L[n; \Delta_k] \tag{10-12}$$

Defining the time-stacked low-time component as

$$\langle s^L[n; \Delta_k]\rangle = \frac{1}{M} \sum_{i=1}^{M} s_i^L[n + m_i; \Delta_k] \tag{10-13}$$

we may write, using equation (10-12),

$$\langle s^L[n; \Delta_k] \rangle = w^L[n; \Delta_k] \star \langle r^L[n; \Delta_k] \rangle \tag{10-14}$$

Given the behavior of the second term in equation (10-14), we conclude that

$$\langle s^L[n; \Delta_k] \rangle \simeq w^L[n; \Delta_k] \tag{10-15}$$

Furthermore, given the essentially low-time characteristics of the (bandpass-mapped) seismic wavelet as previously discussed, we may associate it with the time-stacked low-time component, that is,

$$w[n; \Delta_k] \simeq \langle s^L[n; \Delta_k] \rangle \tag{10-16}$$

Let us illustrate this procedure by short-time-analyzing a time-invariant seismogram resulting from the excitation of the lossless earth structure of Table 10.1 with the seismic airgun signature of Figure 10.11a. Using the same set of short-time windows illustrated in Figure 10.1, the corresponding low-time estimates $s_i^L[n; \Delta_k]$ are illustrated in Figure 10.10. After time stacking, the pulse of Figure 10.11b is obtained, which compares rather favorably with the original seismic wavelet.

This procedure can, of course, be easily generalized to multichannel seismic analysis, as long as the stationarity of the wavelet shape, within the analysis intervals, is kept throughout the channels. We wish to clearly acknowledge the fact that the method proposed here relies rather heavily on the convergence properties of the stacked low-time component $\langle r^L[n; \Delta_k] \rangle$ toward an impulse.

In the absence of a statistical analysis framework within which such convergence behavior could be studied, there remain two possibilities of enforcing it. The first one, which can perhaps be applied in a fully automatic scheme involving multishot/multichannel/multiwindowing analysis, relies on the law of large numbers to guarantee the desired convergence. The second one is based on an interactive framework, within which the human operator is called upon to judge the quality of each individual low-time component, prior to stacking, and to discard those which

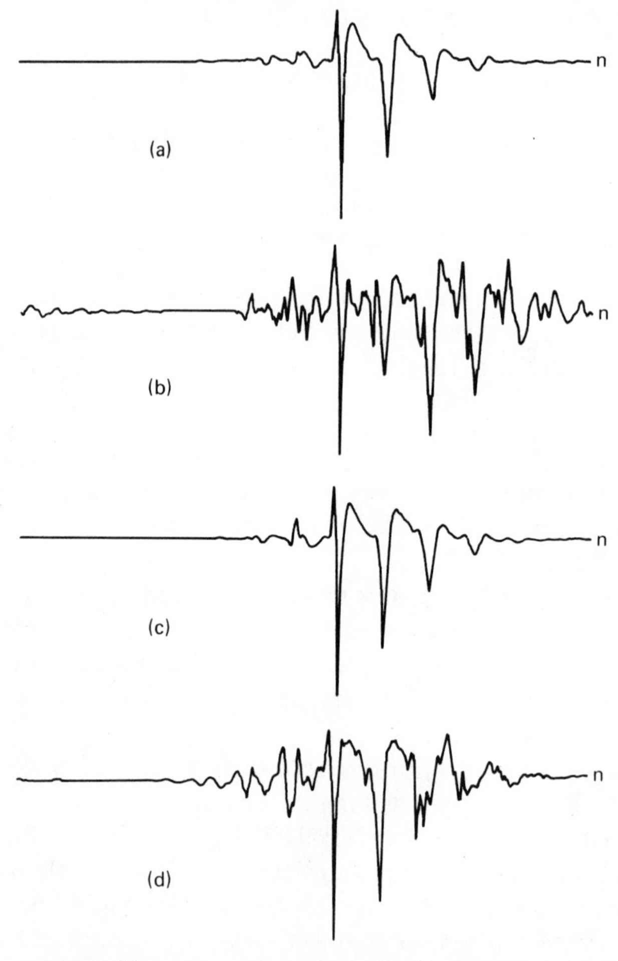

Figure 10.10 Short-time analysis of a synthetic seismogram: (a) low-time component of Hamming windowed section $s_1^L[n; \Delta_k]$; (b) low-time component of exponentially windowed section $s_2^L[n; \Delta_k]$; (c) low-time component of Rayleigh windowed section $s_3^L[n; \Delta_k]$; (d) low-time component of Gaussian windowed section $s_4^L[n; \Delta_k]$

are thought to contain unusually high convolutional noise components.

This has been, in fact, the approach followed in this book. The criteria applied are based on the known structual properties of the signal components: one is a short sequence and the other a

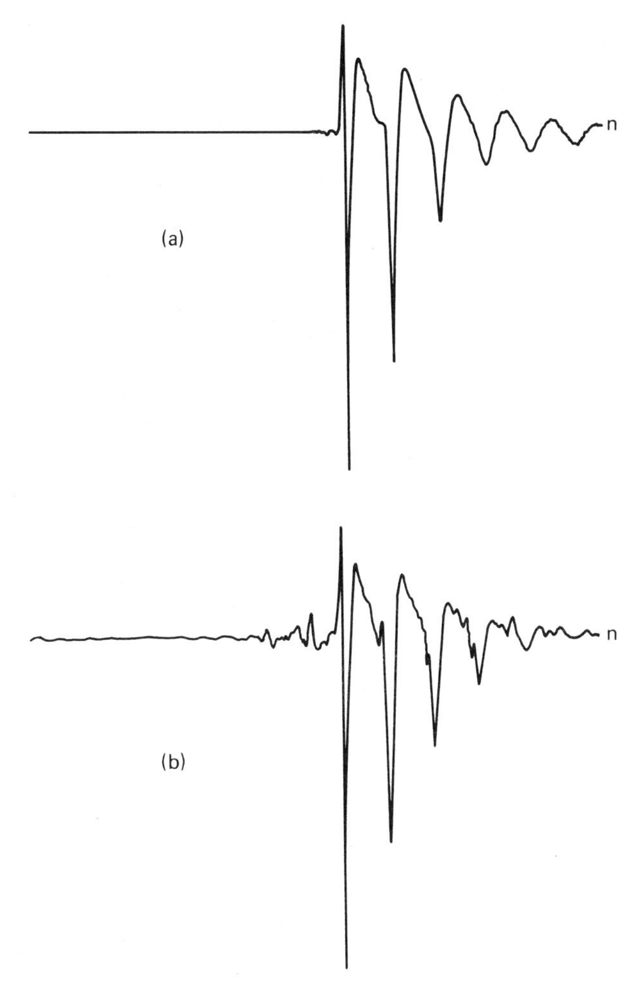

Figure 10.11 Wavelet estimation by homomorphic filtering: (a) original wavelet $w[n; \Delta_k]$; (b) recovered wavelet by time-stacking of low-time components $<s^L[n; \Delta_k]>$

well-organized longer sequence of spiked events. Two basic criteria have then been applied in judging the quality of the low-time analysis. By the first criterion, we discard all low-time estimates that do not exhibit a clear visual correlation with the corresponding short-time segment. By the second criterion we discard all estimates that are unusually long in terms of the expected seismic wavelet length. We shall conclude this chapter by exploring the behavior of this wavelet estimation scheme in the presence of additive noise.

Consider, then, the noisy short-time segment of Figure 10.12a. The signal-to-noise ratio is 10 dB. Two low-time estimates have

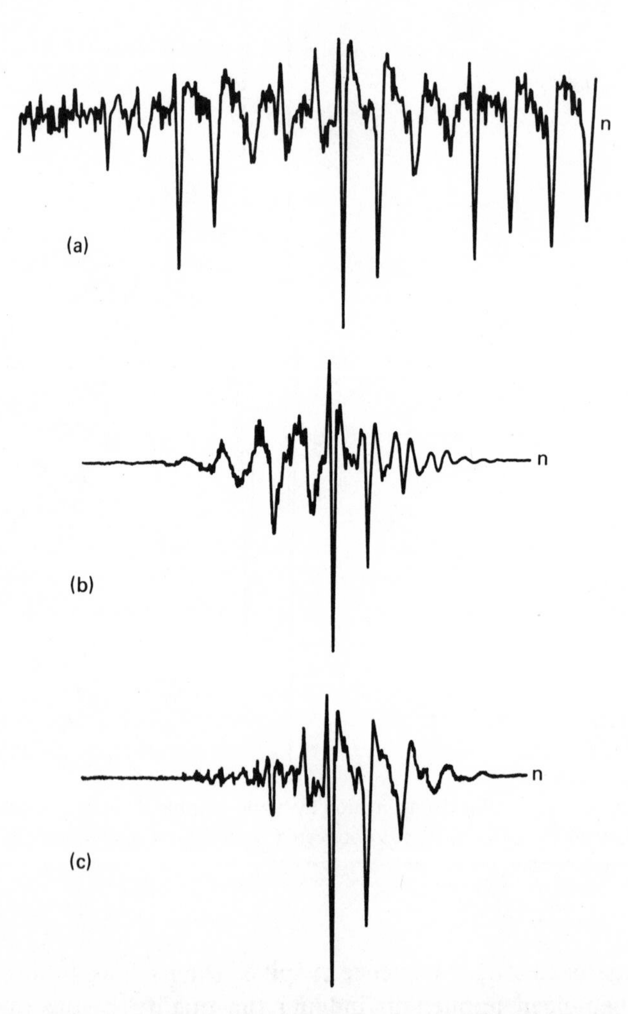

Figure 10.12 Short-time analysis of a noisy synthetic seismogram: (a) trace segment $s[n; \Delta_k]$; (b) low-time component of windowed segment $s_1^L[n; \Delta_k]$; (c) low-time component of windowed segment $s_2^L[n; \Delta_k]$

been computed, using two different windows, and are illustrated in Figures 10.12b and c. As is clear, the corresponding low-time components are now corrupted by noise, but in comparison with the low-time components obtained from the noiseless data shown in Figure 10.10, this seems to be a rather graceful degradation.

This observation matches, in fact, similar observations found in the literature concerning the general robustness of homomorphic systems under additive noise conditions.

X.4 SUMMARY

A strategy for wavelet estimation by short-time homomorphic analysis was devised which capitalizes on the sensitivity of the cepstral structure of mixed-phase, aperiodic impulse trains to time-domain amplitude perturbations. The procedure consists essentially in using different short-time windows on the same seismic segment. Each of these windowed segments is then low-time homomorphically filtered. The seismic wavelet is then estimated by time stacking the resulting sequences, after appropriate synchronization. Alternatively, the low-time cepstral components of the segments may be stacked, yielding an estimate of the cepstrum of the seismic wavelet. These methods may be applied in a fully automatic scheme involving multichannel/multiwindowing, and within the limits of spatial stationarity, also multishot short-time wavelet estimation, since then the total number of elements averaged may be quite large. Single trace analysis requires, in general, human monitoring. Within this context, time stacking yields superior results.

XI

Parametric Deconvolution Using Homomorphic Wavelet Estimation

XI.1 INTRODUCTION

As discussed in Chapter VI, the goal of seismic reflection analysis is the recovery of the reflector series $r[n]$, from which subsequent interpretation in terms of the depths and reflectivities of the subsurface reflecting horizons become possible. To achieve such a goal it is necessary to compress each wavelet $A_k w[n - n_k, \gamma_k]$ into an impulse $A_k \delta[n - n_k]$, while rejecting any additive noise components present in the data. Such an ideal processor has been depicted in Figure 6.5.

Thus far, by taking advantage of the stationarity of the wavelet shape within selected short-time intervals Δ_k, we have developed short-time homomorphic analysis strategies for the estimation of the wavelet shape within an interval Δ_k, $w[n; \Delta_k] = w[n; \gamma_k]$, where γ_k represents the lumped attenuation coefficient associated with the arrivals within Δ_k. Therefore, on a short-time basis, the wavelet compression problem can be formulated as an *inverse filtering* problem; that is, the recovery of the reflector information is accomplished by filtering each data segment through a filter that attempts to transform the wavelet pulse $w[n; \Delta_k]$ into an impulse. Of course, when defining such an inverse filter for wavelet compression, one must take into account the presence of additive noise in both the data and in the wavelet estimate. Therefore, it is usually not appropriate to remove the wavelet effects from a short-time segment $s[n; \Delta_k]$ by simply dividing its Fourier transform by the Fourier transform of the wavelet estimate. Or, equivalently, the estimation of the reflector segment cepstrum $\hat{r}[n; \Delta_k]$ by subtracting the cepstra of the wavelet estimate from the cepstra of the data usually leads to rather unreliable reflector estimates, since no noise-rejection mechanism is implicit in such an approach.

The use of high-time cepstral gating as a means of recovering the reflector series must also be rejected, as repeatedly pointed out in preceding chapters. One alternative is the use of parametric inverse filter design techniques, since these techniques often provide robust representations for signals matched to their basic parametric models.

In this chapter we explore two parametric deconvolution techniques, both based on the combination of the short-time homomorphic wavelet estimation schemes devised in Chapter X, with parametric inverse filtering techniques. The first method to be discussed uses optimum-lag Wiener filtering for the removal of the wavelet effects. This is a straightforward approach in that all the inverse filtering design effort is assigned to the parametric modeling of the wavelet estimate. It is, however, a rather powerful deconvolution algorithm, based on ease of design and the well-known robustness of Wiener filters.

The second method explores homomorphic prediction [6] as a basis for the design of the inverse filter. Homomorphic prediction denotes a combination of homomorphic signal analysis with linear predictive signal modeling. The role that homomorphic filtering is called to play within this context is quite *distinct* from the role it plays within the wavelet estimation context. The basic idea here is to use homomorphic filtering to transform the wavelet estimate into one or more other signals, whose structures are consistent with the basic linear predictive models used.

Both parametric deconvolution algorithms have been tested on synthetic data. As pointed out in Chapter I, the use of synthetic data as a vehicle for the qualitative evaluation of the algorithms proposed has the unique advantage of allowing unequivocal interpretation of the reflector-series estimates in terms of the earth structure. To render the data more realistic, both frequency-dependent earth attenuation and additive noise have been incorporated in the data models.

The results obtained with the parametric deconvolution techniques discussed here were very encouraging. The ultimate test of these algorithms will be, of course, their performance in the face of actual data. This constitutes an area for future research.

XI.2 PARAMETRIC DECONVOLUTION USING WIENER INVERSE FILTERING

Consider the parametric deconvolution strategy illustrated in Figure 11.1. This strategy consists of:

1. The use of homomorphic filtering for wavelet estimation.
2. The design of an optimum-lag digital Wiener filter. This class of filters has been discussed in Chapter VII. In general, it is possible to account for the presence of additive noise in the data by trading off resolution with noise rejection. More details on this subject can be found in [35].
3. The estimation of the reflector series by filtering the seismic data through the Wiener inverse filter.

We illustrate next the use of this strategy in the analysis of a synthetic seismogram. The data were generated assuming the earth structure of Table 11.1.

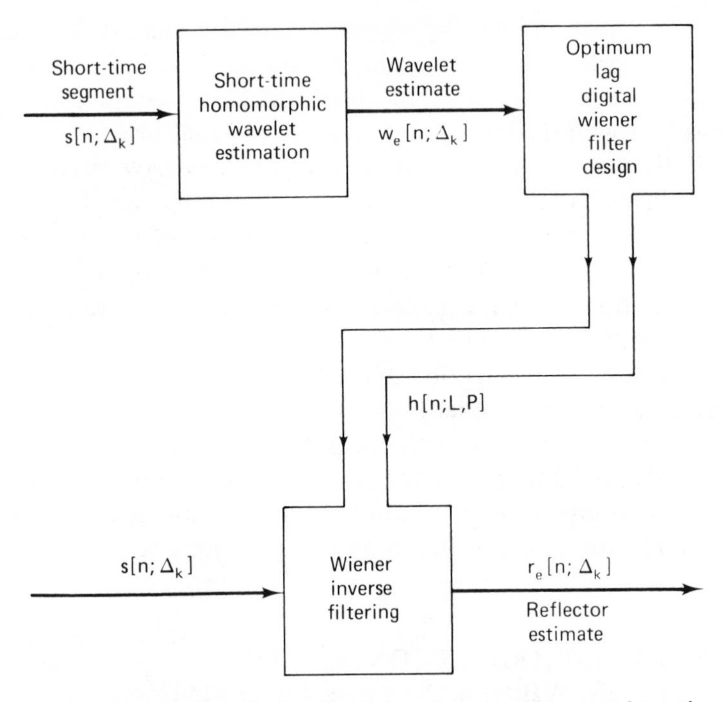

Figure 11.1 Parametric deconvolution: homomorphic wavelet estimation and Wiener inverse filtering

This earth model is similar to that of Table 10.1, except that it incorporates frequency-dependent attenuation, leading to the change of wavelet shape as it travels through the various layers.

The reflector series $r[n]$ associated with this earth model is therefore the same as in Figure 10.1a. We shall illustrate the analysis of these data on the short-time interval Δ_k depicted in Figure 10.1.

Noise was added to the data, with a signal-to-noise ratio of 20 dB. The corresponding seismic segment was low-pass-filtered and bandpass mapped to half of the original frequency band, thus corresponding to a 2:1 decimation. This segment, $\tilde{s}[n; \tilde{\Delta}_k]$, is illustrated in Figure 11.2a. Using a Hamming, a Hamming squared, and a Rayleigh window, the time-stacking homomorphic wavelet

TABLE 11.1

layer number	*two-way travel time within layer (ms)*	*reflection coefficient at top interface*	*attenuation coefficient*
1	111	0.30	0.10×10^{-3}
2	302	0.10	0.15×10^{-3}
3	100	−0.03	0.20×10^{-3}
4	503	0.05	0.25×10^{-3}
5	154	0.03	0.30×10^{-3}
6	219	0.08	0.35×10^{-3}
7	138	−0.04	0.40×10^{-3}
8	286	0.12	0.45×10^{-3}
9	167	0.08	0.50×10^{-3}
10	431	0.06	0.55×10^{-3}
11	∞	0.15	

estimation scheme discussed in Chapter X was applied, yielding the wavelet estimate $\tilde{w}_e[n; \tilde{\Delta}_k]$, depicted in Figure 11.2b.

From such an estimate, an optimum-lag Wiener filter of length 25 was designed. This choice of filter length is appropriate for the compression of the bubble-pulse shape into an impulse. The bubble-pulse reverberation train, however, having an all-zero z-transform, cannot be properly matched by the all-zero structure of the Wiener inverse filter unless the filter length is greatly increased. The use of very large inverse filter lengths must in general be avoided, since then the filter starts to capture all the fine structure of the wavelet estimate. By filtering the segment $\tilde{s}[n; \tilde{\Delta}_k]$ through this filter, the output $\tilde{r}_e[n; \tilde{\Delta}_k]$ of Figure 11.3b was obtained. Figure 11.3a depicts the bandpass-mapped reflector segment $\tilde{r}[n; \tilde{\Delta}_k]$.

The comparison between the reflector segment and its estimate shows the effective power of this approach in handling the noisy characteristics of both the data segment and of the particular wavelet estimate used and in compressing the wavelet bubble pulse. Further removal of the bubble-pulse reverberation effects

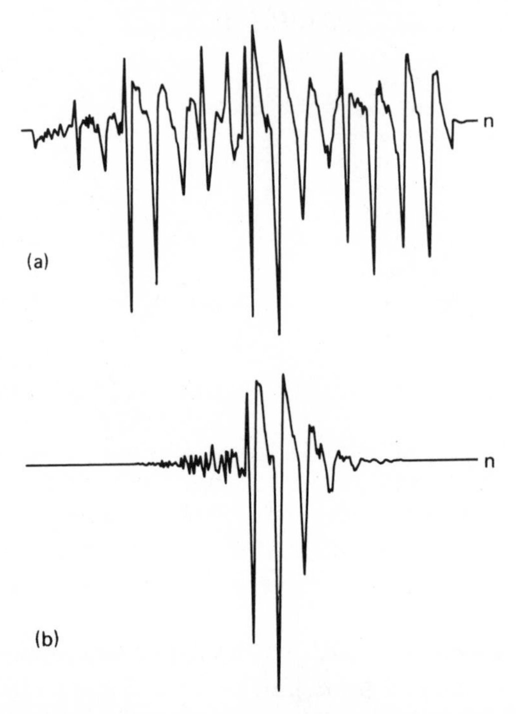

Figure 11.2 Short-time homomorphic wavelet estimation: (a) short-time segment $\tilde{s}[n; \tilde{\Delta}_k]$ (2:1 decimation); (b) wavelet estimate $\tilde{w}_e[n; \tilde{\Delta}_k]$

could be attempted through predictive deconvolution with a prediction distance matched to the bubble-pulse periodicity, since, as discussed in Chapter VII, these filters are especially suited for dereverberation.

XI.3 PARAMETRIC DECONVOLUTION USING HOMOMORPHIC PREDICTION

Homomorphic prediction [6] denotes a signal analysis philosophy characterized by the combination of homomorphic filtering with linear predictive modeling. Homomorphic prediction has been explored as an inverse filter design technique for mixed-phase wavelets [49] and as a technique for pole–zero signal model-

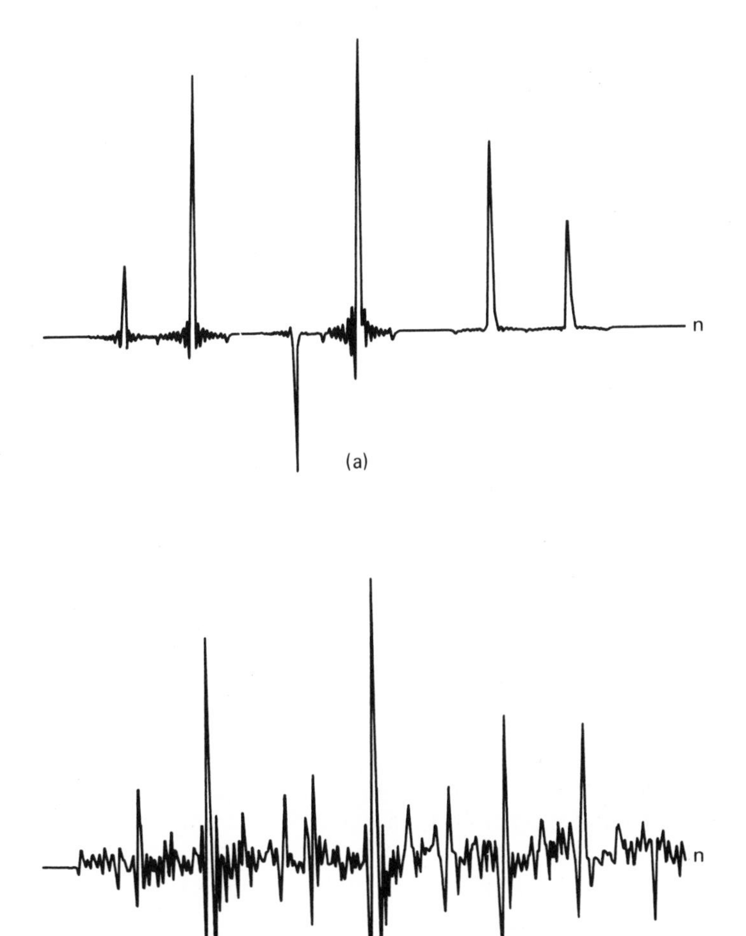

Figure 11.3 Short-time parametric deconvolution using optimum-lag Wiener inverse filtering: (a) reflector segment $\tilde{r}[n; \tilde{\Delta}_k]$; (b) estimated reflector segment $\tilde{r}_e[n; \tilde{\Delta}_k]$

ing [50]. The basic strategy for inverse filter design consists of analyzing the wavelet estimate into components that can be well represented in terms of linear predictive models. For example, one may decompose the wavelet estimate into its minimum-phase and maximum-phase components. As discussed in Chapter VII, the class of causal linear predictive filters having transfer functions of the form

$$H_c(z) = A_c\left(1 - \sum_{k=1}^{M_c} a_k z^{-k}\right)$$

is appropriate for minimum-phase wavelet compression. These filters are minimum-phase.

It can be similarly shown that the class of anticausal predictive filters having transfer functions of the form

$$H_{ac}(z) = A_{ac}\left(1 - \sum_{k=1}^{M_{ac}} b_k z^{+k}\right)$$

are appropriate for maximum-phase wavelet compression. These filters are maximum-phase.

Given these facts, we were originally motivated in [49] to synthesize the inverse filters in terms of a cascade of causal and anticausal linear predictive filters, as illustrated in Figure 11.4. This approach capitalizes on the fact that, as the lengths of the filters tend to infinity, they will tend to the inverses of the minimum-phase and maximum-phase wavelet components. The cascade of these filters will then equal, in the limit, the inverse of the mixed-phase wavelet.

We next illustrate this approach by analyzing the same short-time segment previously used in this chapter, except that this time the bandpass mapping applied to the synthetic seismogram corresponds to a 4:1 decimation. The resulting segment $\tilde{s}[n; \tilde{\Delta}_k]$ is depicted in Figure 11.5. The bandpass-mapped reflector segment $\tilde{r}[n; \tilde{\Delta}_k]$ is depicted in Figure 11.8a. The short-time wavelet estimation by homomorphic filtering was based this time on the use of three Gaussian windows. The resulting time-stacked estimate $\tilde{w}_e[n; \tilde{\Delta}_k]$ was then Hamming-windowed prior to minimum-phase/maximum-phase decomposition. The wavelet estimate and its components are illustrated in Figures 11.6b, c, and d.

The windowing of the wavelet estimate has proved to be very helpful in controlling the spectral dynamic range of the minimum-phase and maximum-phase wavelet components. As we pointed

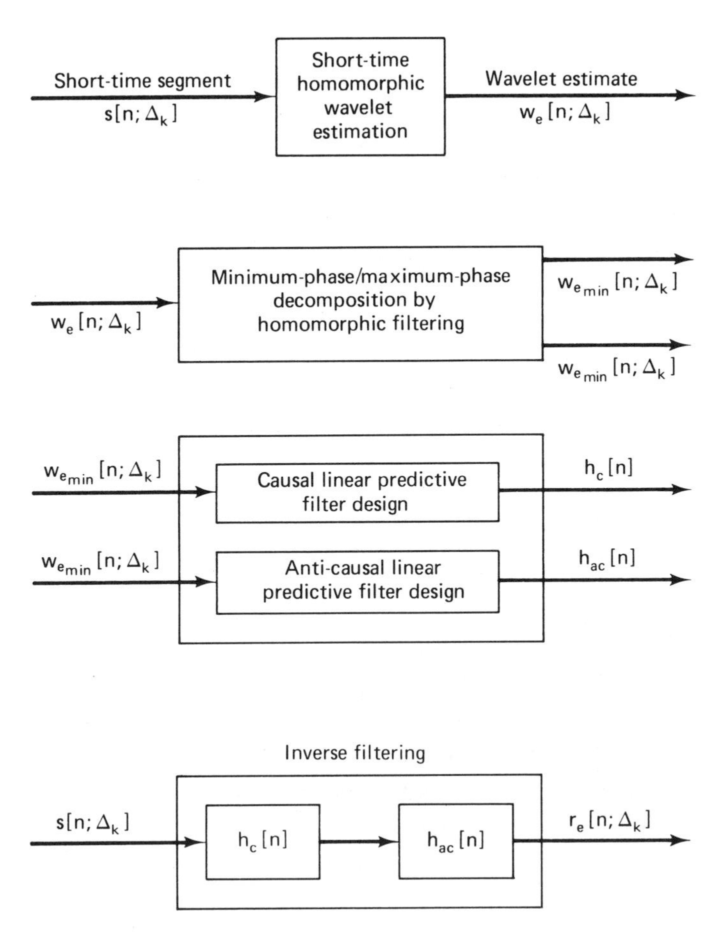

Figure 11.4 Parametric deconvolution: homomorphic wavelet estimation and minimum-phase/maximum-phase homomorphic predictive inverse filtering

out before, we wish to analyze the wavelet estimate into components that may be well represented by linear predictive models. One important aspect of linear prediction is its spectral-matching properties [51]. In particular, for a given filter length, the representation of a signal depends on the smoothness and the dynamic range of its spectrum. However, it may often happen that a mixed-phase signal with a certain spectral dynamic range, say 60 dB, results from the convolution of minimum-phase and maximum-

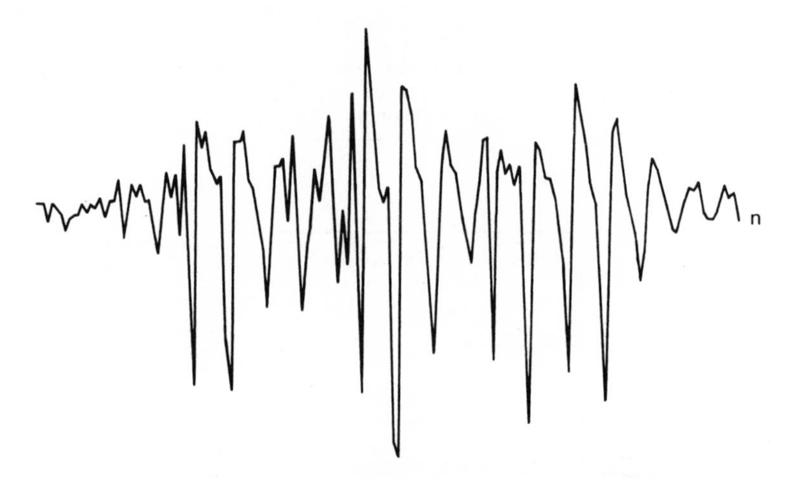

Figure 11.5 Short-time segment $\tilde{s}[n; \tilde{\Delta}_k]$ (4 : 1 decimation)

phase components, each having much larger dynamic ranges. The windowing of the wavelet estimate smooths some of the very sharp zeros that may be present in its spectrum. This has been found to be a most effective means of controlling the dynamic range of the minimum-phase and maximum-phase components, thus allowing better causal and anticausal linear predictive representations. The log spectral magnitudes of the windowed estimate and its components are shown in Figure 11.7.

Another important consideration in linear predictive modeling is that it is matched to the representation of all-pole signals. If a signal is not all-pole but its zeros are broadband, it is often possible to obtain a good representation of such signals with a relatively short filter length. However, the representation of a signal with very sharp zeros requires very large filter lengths. The observation of the spectral structure of the minimum-phase component $\tilde{w}_{e_{\min}}[n; \tilde{\Delta}_k]$ indicates that this component is well matched to linear prediction inverse filtering, since it is essentially characterized by the wavelet resonances. On the contrary, the maximum-phase wavelet component is characterized by the presence of antiresonances, some with very narrow bandwidths. It is then clear that the use of a reasonably short anticausal predictive filter will not lead to an effective compression of this component. This,

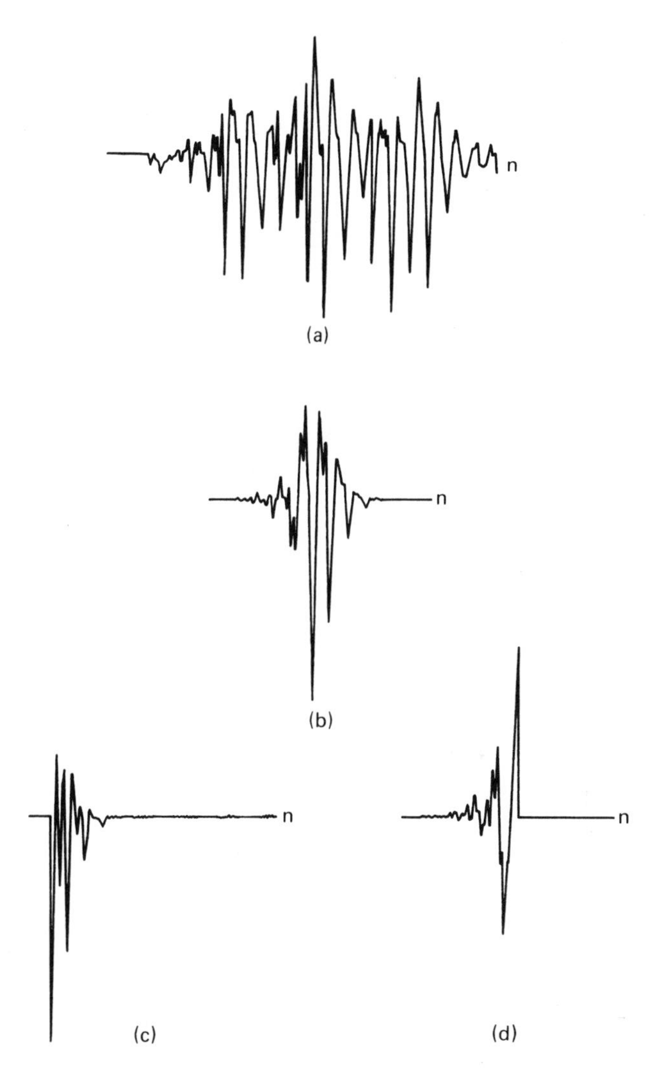

Figure 11.6 Short-time wavelet estimation and minimum-phase/maximum-phase wavelet decomposition: (a) short-time segment $\tilde{s}[n; \tilde{\Delta}_k]$; (b) wavelet estimate $\tilde{w}_e[n; \tilde{\Delta}_k]$; (c) minimum-phase component $\tilde{w}_{e_{\min}}[n; \tilde{\Delta}_k]$; (d) maximum-phase component $\tilde{w}_{e_{\max}}[n; \tilde{\Delta}_k]$

however, is not a problem, since this component is very spiked anyway.

Figure 11.8 illustrates, then, the deconvolution of the segment $\tilde{s}[n; \tilde{\Delta}_k]$ of Figure 11.5, using homomorphic predictive inverse

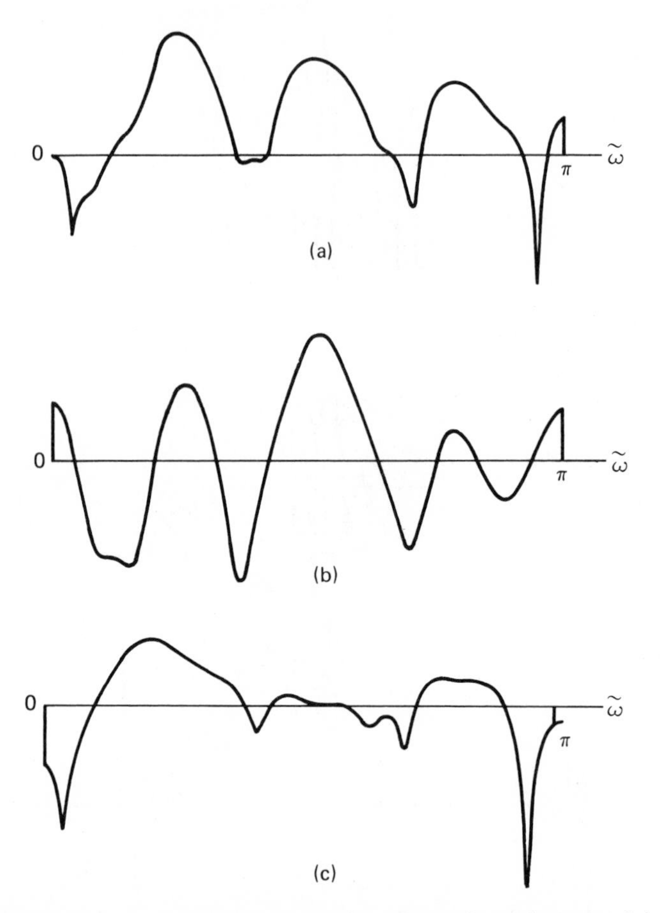

Figure 11.7 Log spectral magnitudes of wavelet estimate and its minimum- and maximum-phase components: (a) $\log|\tilde{W}_e(e^{j\tilde{\omega}}; \tilde{\Delta}_k)|$; (b) $\log|\tilde{W}_{e_{\min}}(e^{j\tilde{\omega}}; \tilde{\Delta}_k)|$; (c) $\log|\tilde{W}_{e_{\max}}(e^{j\tilde{\omega}}; \tilde{\Delta}_k)|$

filtering. In Figure 11.8b, the response of the causal predictive inverse filter $h_c[n]$ to $\tilde{s}[n; \tilde{\Delta}_k]$ is shown. The filter length used corresponds to $M_c = 10$. As is clear from the figure, and for reasons similar to those previously discussed in this chapter in the context of Wiener inverse filtering, this filter did not compress the bubble-pulse reverberation train, nor was it supposed to. Of course, since the reverberation train operator is minimum-phase, it is not represented in the maximum-phase wavelet component. As a consequence, the use of the anticausal filter $h_{ac}[n]$ will not

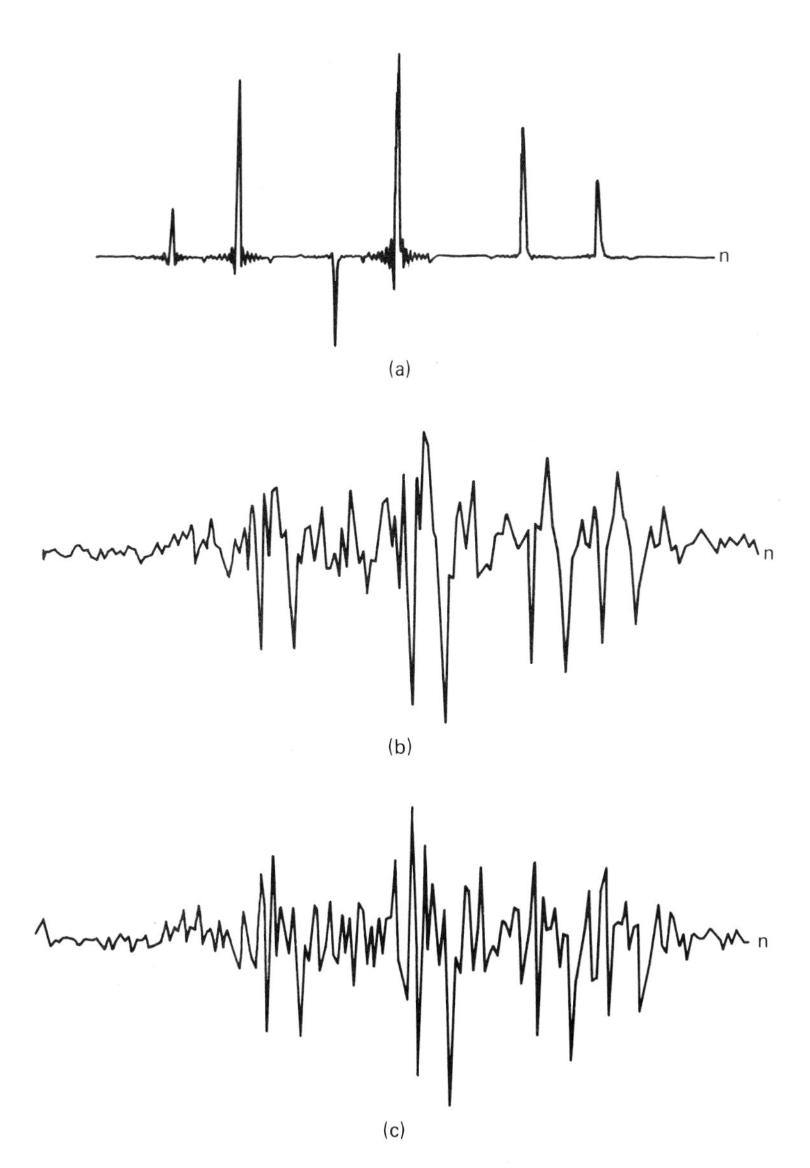

Figure 11.8 Short-time parametric deconvolution using homomorphic prediction: (a) reflector segment $\tilde{r}[n;\tilde{\Delta}_k]$; (b) response of $h_c[n]$ to $\tilde{s}[n;\tilde{\Delta}_k]$; (c) response of $h_c[n] \star h_{ac}[n]$ to $\tilde{s}[n;\tilde{\Delta}_k]$

remove the effects of reverberation, as illustrated in Figure 11.8c, which depicts the output of $h_{ac}[n]$ with $M_{ac} = 5$, when the input is the sequence of Figure 11.8b.

As discussed in the preceding section, one may attempt to remove the bubble-pulse phenomena by using a predictive decon-

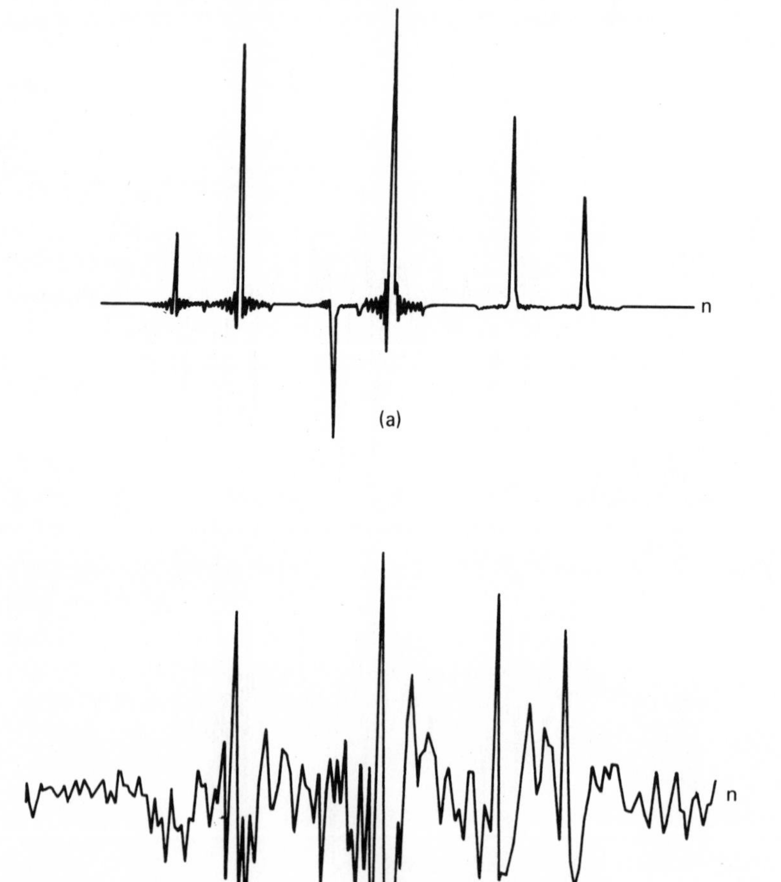

Figure 11.9 Short-time parametric deconvolution using homomorphic prediction: (a) reflector segment $\tilde{r}[n; \tilde{\Delta}_k]$; (b) response of $h_c[n] \star h_{PD}[n]$ to $\tilde{s}[n; \tilde{\Delta}_k]$

volution filter, with a prediction distance matched to the bubble-pulse periodicity.

We designed such a predictive deconvolution filter from the minimum-phase component of the wavelet estimate. By filtering the output of $h_c[n]$ through such a predictive deconvolution filter $h_{PD}[n]$, the reflector estimate $\tilde{r}_e[n; \tilde{\Delta}_k]$ of Figure 11.9b is obtained, depicting a remarkable increase in the resolution of the various arrivals.

Homomorphic prediction is, at the present and in the context of parametric deconvolution, a somewhat speculative technique. The reference to homomorphic prediction in this book was motivated by the increase in inverse filter design flexibility that is associated with this analysis philosophy and has the purpose of encouraging further research in this topic. The examples shown above are simply meant to be illustrative of a potential to be investigated further.

XI.4 SUMMARY

Two parametric deconvolution techniques using homomorphic wavelet estimation were discussed in this chapter. The first technique represents a straightforward combination of homomorphic wavelet estimation with optimum-lag Wiener inverse filter design. The second relies on the use of homomorphic prediction for the design of an inverse filter for the wavelet estimate. The use of both techniques was demonstrated on synthetic, noisy, time-varying seismograms.

XII

Conclusions

The purpose of this book has been to explore the use of homomorphic systems for seismic signal analysis.

The careful study of the homomorphic analysis methods previously employed by a number of researchers to deconvolve seismic reflection and teleseismic data revealed a number of problems, involving: (1) the class of characteristic systems used, (2) the phase-unwrapping algorithms employed, and (3) the analysis strategies followed.

The problems associated with the class of characteristic systems used are due to the fact that such a class of systems is only appropriate for the analysis of full-band signals. Given the bandpass characteristics of seismic data, the use of full-band homomorphic systems is not only theoretically incorrect but is practically ill-advised, since it relies on the presence of the out-of-band noise, leading to nonrobust analysis.

One of the principal contributions of this book was to realize the importance of the problem described above and to solve this problem by introducing a new class of characteristic system which is matched to the analysis of bandpass-filtered signals.

The problem associated with the phase-unwrapping algorithms employed in the past have been reported by many authors. Another principal contribution consists of the careful investigation of these problems and the development of an adaptive phase-unwrapping algorithm which has proved to be very reliable.

All the previous applications of homomorphic signal processing to seismic data analysis published in the literature share a common analysis framework, characterized by the representation of seismic traces in terms of time-invariant models and the use of exponential weighting, when necessary, to ensure the minimum-phase character of the seismic reflector series.

A third important contribution consists of demonstrating that this analysis strategy introduces deterministic constraints on the estimation of the seismic wavelet. In fact, the low-time values of the trace's complex cepstrum are uniquely determined by the first wavelet arrival and depend in no way on subsequent arrivals. This fact is a major limitation to the use of any technique built upon such an analysis framework.

We have proposed a new analysis strategy for homomorphic processing of seismic data. By taking advantage of the slow-time variation of the seismic wavelet characteristics, short-time signal models are developed that represent the seismic trace as a series of windowed time-invariant models. The homomorphic analysis of such short-time segments implies an understanding of the distortion effects caused by the windowing operation.

It was shown that the seismic wavelet is represented in the low-time region of the segment's complex cepstrum, as long as the windowed reflector segment is mixed-phase. Using this fact, strategies for short-time homomorphic wavelet estimation were developed which are based on the use of different short-time windows on the same seismic segment. Each of these windowed segments is then low-time homomorphically filtered and stacked either in the cepstral or in the time domain, yielding an estimate of the seismic wavelet. These strategies have been illustrated on the analysis of synthetic, noisy, time-varying seismograms with good results.

Finally, we have shown that the recovery of the seismic reflector series by homomorphic filtering leads to rather unreliable results, since it incorporates no mechanism that accounts for the presence of noise both in the data and in the wavelet estimate. As a consequence, this recovery is accomplished by parametric deconvolution techniques, which combine homomorphic wavelet estimation with parametric inverse filtering such as optimum-lag Wiener filtering or homomorphic predictive inverse filtering.

The ultimate test of the algorithms proposed will be their performance in the face of actual data. We hope that this book demonstrates the potential of homomorphic filtering and guides its future applications to seismic signal analysis.

References

[1] A. V. Oppenheim and R. W. Schafer, *Digital Signal Processing*, Prentice-Hall, Inc., Englewood Cliffs, N.J., 1975.

[2] A. V. Oppenheim, "Superposition in a Class of Nonlinear Systems," *Technical Report No. 432*, Research Laboratory of Electronics, M.I.T., Cambridge, Mass., Mar. 1965.

[3] K. Hoffman and R. Kunze, *Linear Algebra*, Prentice-Hall, Inc., Englewood Cliffs, N.J, 1961.

[4] A. V. Oppenheim and R. W. Schafer, "Homomorphic Analysis of Speech," *IEEE Trans. Audio Electroacoust.*, vol. AU-16, no. 2, pp. 221–226, June 1968.

[5] A. V. Oppenheim, "A Speech Analysis–Synthesis System Based on Homomorphic Filtering," *J. Acoust. Soc. Amer.*, vol. 45, pp. 458–465, Feb. 1969.

[6] A. V. OPPENHEIM, G. KOPEC, and J. TRIBOLET, "Signal Analysis by Homomorphic Prediction," *IEEE Trans. Acoustics, Speech and Signal Processing*, vol. ASSP-24, no. 4, pp. 327–332, Aug. 1976.

[7] R. C. KEMERAIT and D. G. CHILDERS, "Signal Detection and Extraction by Cepstrum Techniques," *IEEE Trans. Inform. Theory*, vol. IT-18, no. 6, pp. 745–759, Nov. 1972.

[8] A. V. OPPENHEIM and J. TRIBOLET, "Application of Homomorphic Filtering to Seismic Data Processing," in *Applied Time Series Analysis*, D. Findley, editor, Academic Press, Inc., New York, 1978.

[9] B. P. BOGERT, M. J. HEALY, and J. W. TUKEY, "The Quefrency Analysis of Time Series for Echoes: Cepstrum, Pseudo-Autocovariance, Cross-Cepstrum, and Saphe Cracking," in M. Rosenblatt (ed.), *Time Series Analysis*, John Wiley & Sons, Inc., New York, pp. 209–243, Chap. 15, 1963.

[10] D. G. CHILDERS, D. P. SKINNER, and R. C. KEMERAIT, "The Cepstrum: A Guide to Processing," *Proc. IEEE*, vol. 65, no. 10, pp. 1428–1443, Oct. 1977.

[11] K. STEIGLITZ and B. DICKINSON, "Computation of the Complex Cepstrum by Factorization of the z-Transform," 1977 IEEE International Conference on Acoustics, Speech and Signal Processing, Hartford, Conn., May 9–11, 1977.

[12] R. W. SCHAFER, "Echo Removal by Discrete Generalized Linear Filtering," *Technical Report No. 466*, Research Laboratory of Electronics, M.I.T., Cambridge, Mass., Feb. 1969, also *Ph.D. Thesis*, Department of Electrical Engineering, M.I.T., Cambridge, Mass., Feb. 1968.

[13] J. M. TRIBOLET, "A New Phase Unwrapping Algorithm," *IEEE Trans. Acoustics, Speech and Signal Processing*, vol. 25, no. 2, pp. 170–177, Apr. 1977.

[14] R. E. CROCHIERE and L. R. RABINER, "Optimum FIR Digital Filter Implementation for Decimation, Interpolation and Narrow Band Filtering," *IEEE Trans. Acoustics, Speech and Signal Processing*, vol. 23, no. 5, pp. 444–456, Oct. 1975.

[15] J. H. AHLBERG, E. N. NILSSON, and J. L. WALSH, *The Theory*

of Splines and Their Application, Academic Press, Inc., New York, 1967.

[16] B. BHANU and J. MCCLELLAN, private communications on adaptive phase unwrapping using splines, Digital Signal Processing Group, Research Laboratory of Electronics, M.I.T., Cambridge, Mass., 1977.

[17] L. R. RABINER, J. H. MCCLELLAN, and T. W. PARKS, "FIR Digital Filter Design Techniques Using Weighted Chebyshev Approximation," *Proc. IEEE—Special Issue on Digital Signal Processing*, vol. 63, no. 4, pp. 595–610, Apr. 1975.

[18] A. V. OPPENHEIM, W. F. MECKLENBRÄUKER, and R. M. MERSEREAU, "Variable Cutoff Linear Phase Digital Filters," *IEEE Trans. Circuits and Systems*, vol. CAS-23, no. 4, pp. 199–203, Apr. 1976.

[19] M. DOBRIN, *Introduction to Geophysical Prospecting*, McGraw-Hill Book Company, New York, 1976.

[20] F. S. GRANT and G. F. WEST, *Interpretation Theory in Applied Geophysics*, McGraw-Hill Book Company, New York, 1965.

[21] W. MAYNE, "Common Reflection Point Horizontal Data Stacking Techniques," *Geohpysics*, vol. 27, no. 6, pp. 927–938, Dec. 1962.

[22] L. BERRYMAN et al., "Reflections for Multiple Transition Layers," *Geophysics*, vol. 23, no. 2, pp. 223–243, Apr. 1958.

[23] P. WUENSCHEL, "Seismogram Synthesis Including Multiples and Transmission Coefficients," *Geophysics*, vol. 25, no. 1, pp. 106–129, Feb. 1960.

[24] A. TROREY, "Theoretical Seismograms with Frequency and Depth Dependent Absorption," *Geophysics*, vol. 27, no. 6. pp. 766–785, Dec. 1962.

[25] J. CLAERBOUT, "Synthesis of a Layered Medium from Its Acoustic Transmission Response," *Geophysics*, vol. 33, no. 2, pp. 264–269, Apr. 1968.

[26] D. MIDDELTON and J. WHITTLESEY, "Seismic Models and Deterministic Operators for Marine Reverberation," *Geophysics*, vol. 33, no. 4, pp. 557–583, Aug. 1968.

[27] S. Treitel and E. Robinson, "Seismic Propagation in Layered Media in Terms of Communication Theory," *Geophysics*, vol. 31, no. 1, pp. 17–32, Feb. 1966.

[28] K. B. Theriault, "Optimum Arrival Time in Exploration Seismology," *M.S. and E.E. Thesis*, Department of Electrical Engineering, M.I.T., Cambridge, Mass., Sept. 1974.

[29] N. Ricker, "The Form and Nature of Seismic Waves and the Structure of Seismograms," *Geophysics*, vol. 5, no. 4, pp. 348–366, Oct. 1940.

[30] K. Peacock and S. Treitel, "Predictive Deconvolution: Theory and Practice," *Geophysics*, vol. 34, no. 2, pp. 155–169, Apr. 1969.

[31] N. Ricker, "Wavelet Contraction, Wavelet Expansion, and the Control of Seismic Resolution," *Geophysics*, vol. 18, no. 4, pp. 769–792, Oct. 1953.

[32] G. Wadsworth et al., "Detection of Reflections on Seismic Records by Linear Operators," *Geophysics*, vol. 18, no. 3, pp. 539–586, July 1953.

[33] E. Robinson, "Predictive Decomposition of Seismic Traces," *Geophysics*, vol. 22, no. 4, pp. 767–778, Oct. 1957.

[34] T. Ulrych, "Application of Homomorphic Deconvolution to Seismology," *Geophysics*, vol. 36, no. 4, pp. 650–660, Aug. 1971.

[35] E. Robinson and S. Treitel, "Principles of Digital Wiener Filtering," *Geophys. Prospecting*, vol. 15, no. 3, pp. 311–333, Sept. 1967.

[36] E. Robinson, *Multichannel Time Series Analysis with Digital Computer Programs*, Holden-Day, Inc., San Francisco, 1967.

[37] S. Treitel and E. A. Robinson, "The Design of High-Resolution Digital Filters," *IEEE Trans. Geosci. Electron.*, vol. GE-4, no. 1, June 1966.

[38] A. B. Baggeroer, "Tapped Delay Line Models for the Dereverberation of Deep Water Multiples," *Technical Report WHOI*-73-66, Woods Hole Oceanographic Institution, Woods Hole, Mass., Oct. 1973.

[39] T. Ulrych, O. G. Jensen, R. M. Ellis, and P. G. Sommerville, "Homomorphic Deconvolution of Some Teleseismic Events," *Bull. Seismol. Soc. Amer.*, vol. 62, no. 5, pp. 1253–1265, Mar. 1972.

[40] R. W. Clayton and R. A. Wiggins, "Source Shape Estimation and Deconvolution of Teleseismic Bodywaves," *Geophys. J. Roy. Astron. Soc.*, vol. 47, pp. 151–177, 1977.

[41] P. G. Sommerville, "Time Domain Determination of Earthquake Fault Parameters from Short-Period P-Waves," *Ph.D. Thesis*, Department of Geophysics and Astronomy, University of British Columbia, Vancouver, B.C., Dec. 1975.

[42] P. L. Stoffa, P. Buhl, and G. M. Bryan, "The Application of Homomorphic Deconvolution to Shallow-Water Marine Seismology—Part I: Models; and Part II: Real Data," *Geophysics*, vol. 39, no. 4, pp. 401–426, Aug. 1974.

[43] B. Buttkus, "Homomorphic Filtering, Theory and Practice," *Geophys. Prospecting*, vol. 23, no. 4, p. 712, 1975.

[44] J. B. Gallemore, "A Comparative Evaluation of Two Acoustic Signal Dereverberation Techniques," *S.M. and E.E. Thesis*, Department of Electrical Engineering and Computer Science, M.I.T., Cambridge, Mass., June 1976.

[45] R. M. Otis and R. B. Smith, "Homomorphic Deconvolution by Log Spectral Averaging," *Geophysics*, 1978.

[46] T. G. Stockham, Jr., T. M. Cannon, and R. B. Ingebretsen, "Blind Deconvolution Through Digital Signal Processing," *IEEE Proc.*, vol. 63, no. 4, pp. 678–692, Apr. 1975.

[47] R. W. Clayton and T. J. Ulrych, "A Restoration Method for Impulsive Functions," *IEEE Trans. Inform. Theory*, pp. 262–264, Mar. 1977.

[48] J. Tribolet, T. Quatieri, and A. Oppenheim, "Short-Time Homomorphic Analysis," 1977 IEEE Conference on Acoustics, Speech and Signal Processing, Hartford, Conn., May 9–11, 1977.

[49] J. M. Tribolet, A. V. Oppenheim, and G. E. Kopec, "Deconvolution by Homomorphic Prediction," presented at the

45th Annual International Meeting of the Society of Exploration Geophysicists, Denver, Oct. 12–16, 1975.

[50] G. E. Kopec, A. V. Oppenheim, and J. M. Tribolet, "Speech Analysis by Homomorphic Prediction," *IEEE Trans. Acoustics, Speech and Signal Processing*, vol. ASSP-25, no. 1, Feb. 1977.

[51] J. Makhoul, "Linear Prediction: A Tutorial Review," *Proc. IEEE*, vol. 63, pp. 561–580, Apr. 1975.

Index